EXAM PREP

Fire Instructor I & II

By Dr. Ben A. Hirst, Performance Training Systems

JONES AND BARTLETT PUBLISHERS
Sudbury, Massachusetts
BOSTON TORONTO LONDON SINGAPORE

Jones and Bartlett Publishers
World Headquarters
40 Tall Pine Drive
Sudbury, MA 01776
978-443-5000
www.jbpub.com

International Association of Fire Chiefs
4025 Fair Ridge Drive
Fairfax, VA 22033
www.IAFC.org

Performance Training Systems, Inc.
760 U.S. Highway One, Suite 101
North Palm Beach, FL 33408
www.FireTestBanks.com

Jones and Bartlett Publishers Canada
6339 Ormindale Way
Mississauga, ON L5V 1J2
Canada

Jones and Bartlett Publishers International
Barb House, Barb Mews
London W6 7PA
United Kingdom

Jones and Bartlett's books and products are available through most bookstores and online booksellers. To contact Jones and Bartlett Publishers directly, call 800-832-0034, fax 978-443-8000, or visit our website www.jbpub.com.

Substantial discounts on bulk quantities of Jones and Bartlett's publications are available to corporations, professional associations, and other qualified organizations. For details and specific discount information, contact the special sales department at Jones and Bartlett via the above contact information or send an email to specialsales@jbpub.com.

Editorial Credits

Author: Dr. Ben A. Hirst

Production Credits

Chief Executive Officer: Clayton E. Jones
Chief Operating Officer: Donald W. Jones, Jr.
President: Robert W. Holland, Jr.
V.P., Sales and Marketing: William J. Kane
V.P., Production and Design: Anne Spencer
V.P., Manufacturing and Inventory Control: Therese Bräuer
Publisher, Public Safety Group: Kimberly Brophy
Acquisitions Editor, Fire: William Larkin
Editorial Assistant: Adrienne Zicht
Production Editor: Karen Ferreira
Director of Marketing: Alisha Weisman
Cover Photograph: © Michael Heller / 911 Pictures
Cover Design: Kristin Ohlin
Interior Design: Anne Spencer
Composition: Northeast Compositors
Printing and Binding: Courier Stoughton

ISBN-13: 978-0-7637-2762-8
ISBN-10: 0-7637-2762-8

The procedures in this text are based on the most current recommendations of responsible sources. The publisher and Performance Training Sysyems, Inc. make no guarantees as to, and assume no responsibility for the correctness, sufficiency, or completeness of such information or recommendations. Other or additional safety measures may be required under particular circumstances. This text is intended solely as a guide to the appropriate procedures to be employed when responding to an emergency. It is not intended as a statement of the procedures required in any particular situation, because circumstances can vary widely from one emergency to another. Nor is it intended that this text shall in any way advise firefighting personnel concerning legal authority to perform the activities or procedures discussed. Such local determination should be made only with the aid of legal counsel.

6048
Printed in the United States of America
10 09 08 07 06 10 9 8 7 6 5 4 3 2

CONTENTS

PREFACE

The Fire and Emergency Medical Service faces one of the most challenging periods in its history. Local, state, provincial, national, and international government organizations are under pressure to deliver ever increasing first response services. The events of September 11, 2001, continued threats by terrorist organizations worldwide, and the need to maximize available funds are part of the reason why most Fire and Emergency Medical Service organizations are reinventing their roles.

One of the challenges of reinventing the Fire and Emergency Medical Service roles is increasing professional requirements. Organizations such as the National Fire Protection Association (NFPA), National Professional Qualifications Board (Pro Board), International Fire Service Accreditation Congress (IFSAC), the International Association of Fire Chiefs (IAFC), and the International Association of Fire Fighters (IAFF) have a dramatic influence on raising the professional qualifications of emergency response personnel.

Qualification standards have been improved. Accreditation of training and certification are at the highest levels in the history of the Fire and Emergency Medical Service. These improvements are reflected in a better prepared emergency responder but are not without an effect on those individuals who serve. Fire instructors are now required to expand their roles, acquire new knowledge, develop higher level technical skills, and participate in requalification programs on a regular basis.

The aftermath of September 11, 2001 has had a profound impact on the Fire and Emergency Medical Service. Lessons learned, new technologies, and a national focus on terrorism and weapons of mass destruction are placing much greater demands on fire instructors to change their curriculums. Obvious dangers faced by emergency responders under heightened security conditions require many adjustments in what is being taught and how it must be changed and updated.

Our national leaders are constantly pointing to emergency responders as our "first line of defense" against acts of terror and defense of life and property from dangerous weapons that have not been used extensively in our history. Fire instructors will be the primary delivery system for developing essential knowledge, skills, and abilities of emergency responders to efficiently and effectively meet these awesome responsibilities.

Credentials are even more necessary for fire instructors to advance in their chosen careers. Performance Training Systems, Inc., (PTS) has emerged over the past 18 years to become the leading provider of valid testing materials for certification, promotion, and training for fire and emergency medical personnel. More than 30 examination banks provide the basis for validated examinations. All PTS products are based on the NFPA Professional Qualifications Standards for fire personnel and the Department of Transportation (DOT) Curriculum for emergency medical personnel.

Over the past eight years, PTS has conducted research supporting the development of the Systematic Approach to Examination Preparation (SAEP). The SAEP System results in consistent improvement in examination scores for persons taking certification, promotion, and training completion examinations. This *Exam Prep* series is designed to assist fire instructors to improve their knowledge, skills, and abilities while seeking certification, promotion, and training program completion. Using the features of SAEP, coupled with helpful examination-taking tips and hints, will help ensure an improved performance, and create a more knowledgeable and skilled fire instructor

All examination questions used in SAEP were written by fire and emergency service personnel. Technical content was validated through the use of current technical textbooks and other technical reference materials. Job content was validated by the use of technical review committees representative of the Fire Instructor I and Fire Instructor II ranks and representatives from state fire service certification organizations. The examination questions represent an approximate 50% sample of items from the Fire Instructor I and Fire Instructor II Fire Test Banks developed by PTS over the past 18 years. These examination banks are used by 58 certification agencies in the United States, Canada, and several foreign countries. For more information on the number of available examination banks and the processes of validation visit *www.firetestbanks.com*.

Introduction to the Systematic Approach to Examination Preparation

How does SAEP work? SAEP is an organized process of carefully researched phases that permits each person to proceed through examination preparation at that individual's own pace. At certain points, self-study is required to move from one phase of the program to another. Receiving and then using feedback on one's progress is the basis of SAEP. It is important to follow the program steps carefully to realize the full benefits of the system.

SAEP allows you to prepare carefully for your next comprehensive training, promotional, or certification examination. Just follow the steps to success. PTS, the leader in producing promotional and certification examinations for the Fire and Emergency Medical Service industry for more than 18 years, has both the experience and the testing expertise to help you meet your professional goals.

Using the Exam Prep manual will enable you to pinpoint areas of weakness in terms of NFPA Standard 1041, and the feedback will provide the reference and page numbers to help you research the questions that you miss or guess using current technical reference materials. This program comprises a three-examination set for Fire Instructor I and Fire Instructor II as described in NFPA 1041, *Standard for Fire Instructor Professional Qualifications, 2002 Edition.*

Primary benefits of SAEP in preparing for these examinations include the following:

- Emphasis on areas of weakness
- Immediate feedback
- Savings in time and energy
- Learning technical material through context and association
- Helpful examination preparation practices and hints

SAEP is organized in four distinct phases for Fire Instructor I and Fire Instructor II as described in NFPA 1041, *Standard for Fire Instructor Professional Qualifications, 2002 Edition*. These phases are briefly described next.

Phases of SAEP

SAEP is organized in four distinct phases for Fire Instructor I and Fire Instructor II. The phases are briefly described below.

Phase I

Phase I includes three examinations containing items that are selected from each major part of NFPA 1041 *Standard for Fire Instructor Professional Qualifications, 2002 Edition.*

An essential part of the SAEP design is to survey your present level of knowledge and build on it for subsequent examination and self-directed study activities. Therefore, it is suggested that you read the reference materials but do not study or look up any answers while taking the initial examination. Upon completion of the initial examination, you will complete a feedback activity and record examination items that you missed or that you guessed. We ask you to perform certain tasks during the feedback activity. Once you have completed the initial examination and researched the answers for any questions you missed, you may proceed to the next examination. This process is repeated through and including the third examination in the Fire Instructor I and Fire Instructor II series, depending on the level of certification you are seeking.

Phase II

Fire Instructor II examinations are provided for use in Phase II of SAEP. This phase includes three examinations, each made up of examination items from appropriate sections of NFPA 1041, *Standard for Fire Instructor Professional Qualifications, 2002 Edition*.

The examinations should be completed as prescribed in the directions supplied with the examination. Complete the feedback report using the procedures provided in the answer and feedback section. Pay particular attention to those references covering material on which you score the lowest. At this point, it is important to read the materials containing the correct response in context once again. This technique will help you master the material, relate it to other important information, and retain knowledge.

Phase III

Phase III contains important information about examination-item construction. It provides insight regarding the examination-item developers, the way in which they apply their technology, and hints and tips to help you score higher on any examination. Make sure you read this phase carefully. It is a good practice to read it twice, and study the information a day or two prior to your scheduled examination.

Phase IV

Phase IV information addresses the mental and physical aspects of examination preparation. By all means, do not skip this part of your preparation. Points can be lost if you are not ready-both physically and mentally-for the examination. If you have participated in sporting or other competitive events, you know the importance of this level of preparation. There is no substitute for readiness. Just being able to answer the questions will not help you achieve a level of excellence and move you to the top of the examination list for training, promotion, or certification. Quality preparation involves much more than simply answering examination items.

Supplemental Practice Examination Program

The supplemental practice examination program differs from the SAEP program in several ways. First, it is provided over the Internet 24 hours a day, 7 days a week. In addition, this supplemental practice examination allows you to make final preparations immediately before your examination date. You will receive an immediate feedback report that includes the questions missed and the references and page numbers pertaining to those missed questions. The practice examination can help you concentrate on your areas of greatest weakness and will save you time and energy immediately before the examination date. If you choose this method of preparation, do not "cram" for the examination. The upcoming helpful hints for examination preparation will explain the reasons for avoiding a "cramming exercise." A supplemental practice examination is available with the purchase of this Exam Prep manual by using the enclosed registration form. Do not forget to fax a copy of your Personal Progress Plotter along with your registration form. The data supplied on your Personal Progress Plotter will be kept confidential and will be used by PTS to make future improvements in the Exam Prep series. You may take a short practice examination to get the procedure clear in your mind by going to www.webtesting.cc.

Good luck in your efforts to improve your knowledge and skills. Our ultimate goal is to improve the Fire and Emergency Medical Service one person at a time. We want your feedback and impressions of the system to help us implement improvements in future editions of the Exam Prep series of books. Address your comments and suggestions to www.firetestbanks.com.

Rule 1

Examination preparation is not easy. Preparation is 95% perspiration and 5% inspiration.

Rule 2

Follow the steps very carefully. Do not try to reinvent or shortcut the system. It really works just as it was designed to!

Personal Progress Plotter

Fire Instructor I Exam Prep:

Name: ______________________________

Date Started: ______________________________

Date Completed: ______________________________

Fire Instructor II Exam Prep:

Name: ______________________________

Date Started: ______________________________

Date Completed: ______________________________

Fire Instructor I	Number Guessed	Number Missed	Examination Score
Examination I-1			
Examination I-2			
Examination I-3			

Fire Instructor II	Number Guessed	Number Missed	Examination Score
Examination II-1			
Examination II-2			
Examination II-3			

Formula to compute Examination Score = ((Number guessed + Number missed) × Point value per examination item) subtracted from 100

Note: 100-Item Examination = 1.0 point per examination item

75-Item Examination = 1.34 points per examination item

Example: Examination I-1, 5 examination items were guessed, 8 were missed for a total of 13 on a 100-item examination. The examination score would be 100 − (13 × 1. 0 Points) = 87

Example: Examination II-1, 5 examination items were guessed, 8 were missed for a total of 13 on a 75-item examination. The examination score would be 100 − (13 × 1. 34 points) = 82

Note: To receive your free online practice examination, you must fax a copy of your completed Personal Progress Plotter along with your registration form.

PHASE I

Fire Instructor I

Examination I-1, Beginning NFPA Standard 1041

Examination I-1 contains 75 examination items. Read the reference materials but do not study prior to taking the examination. This examination is designed to identify your weakest areas in terms of NFPA Standard 1041. Some steps in SAEP will require self-study of specific reference materials. Remove Examination I-1 from the book. Mark all answers in ink to ensure that no changes are made later. Do not mark through answers or change answers in any way once you have selected your answers.

Step 1—Take Examination I-1. When you have completed Examination I-1, go to Appendix A and compare your answers with the correct answers. Notice that each answer has reference materials with page numbers. If you missed the correct answer to the examination item, you have a source for conducting your correct answer research.

Step 2—Score Examination I-1. How many examination items did you miss? Write the number of missed examination items in the blank in ink. ________________ Enter the number of examination items you guessed in this blank. ________________ Enter these numbers in the designated locations on your Personal Progress Plotter.

Step 3—Now the learning begins! Carefully research the page cited in the reference material for the correct answer. For instance, if you are using IFSTA, *Fire and Emergency Services Instructor, Sixth Edition*, go to the page number provided and find the answer.

Rule 3

Mark with an "X" any examination items for which you guessed the answer. For maximum return on effort, you should also research any answer that you guessed, even if you guessed correctly. Find the correct answer, highlight it, and then read the entire paragraph that contains the answer. Be honest and mark all questions you guessed. Some examinations have a correction for guessing built into the scoring process. The correction for guessing can reduce your final examination score. If you are guessing, you are not mastering the material.

Rule 4

Read questions twice if you have any misunderstanding, especially if the question contains complex directions or activities

Helpful Hint

Most of the time your first impression is the best. More than 41% of answers changed during our SAEP field test were changed from a right answer to a wrong answer. Another 33% changed their answer from a wrong answer to another wrong answer. Only 26% of changed answers were from wrong to right. In fact, three participants did not make a perfect score of 100% because they changed one right answer to a wrong one! Think twice before you change your answer. The odds are not in your favor.

Helpful Hint

Researching correct answers is one of the most important activities in the SAEP. Locate the correct answer for any item answered incorrectly. Highlight the correct answer. Then read the entire paragraph containing the answer. This will put the answer in context for you and provide important learning by association.

Helpful Hint

Proceed through all missed examination items using the same technique. Reading the entire paragraph improves retention of the information and helps you develop an association with the material and learn the correct answers. This step may sound simple. A major finding during the development and field testing of SAEP was that you learn from your mistakes.

Examination I-1

Directions

Remove Examination I-1 from the manual. First, take a careful look at the examination. There should be 75 examination items. Notice that a blank line precedes each examination item number. This line is provided for you to enter the answer to the examination item. Write the answer in ink. Remember the rule about not changing your answers. Our research has shown that changed answers are often incorrect, and, more often than not, the answer that is chosen first is correct.

If you guess the answer to a question, place an "X" or a check mark by your answer. This step is vitally important as you gain and master knowledge. We will explain how we treat the "guessed" items later in SAEP.

Take the examination. Once you complete it, go to Appendix A and score your examination. Once the examination is scored, carefully follow the directions for feedback on the missed and guessed examination items.

_____ **1.** Performance standards for fire instructors are identified in:

A. NFPA 1021.
B. NFPA 1041.
C. NFPA 1403.
D. NFPA 1031.

_____ **2.** The primary role of the instructor is:

A. planning and conducting training.
B. pursuing knowledge to adapt changes in life and work.
C. reviewing changes in regional curricula.
D. acting as a mentor to aspiring instructors.

_____ **3.** The **primary** purpose of collecting training data and analyzing the information is to:

A. adjust evaluation techniques.
B. meet State Fire Marshal requirements.
C. enhance learning efficiency.
D. provide information to NFIRS.

_____ **4.** Training records will do all of the following **except**:

A. identify areas that need more attention.
B. document training that has been completed.
C. provide justifications for purchase of fireground equipment.
D. be used in a legal case.

_____ **5.** An accident investigation report should document:

A. fault(s).
B. improper attitudes.
C. facts.
D. A and C only

_____ **6.** Analyzing circumstances surrounding accidents can enable an instructor to:

A. identify and locate principal sources of accidents.
B. disclose the nature and size of accident problems in different operations.
C. indicate the need for engineering revisions by identifying unsafe conditions of various types of equipment.
D. All of the above

_____ **7.** A person who performs at the Instructor I level is responsible for:

A. developing lesson plans.
B. coordinating other instructors.
C. presenting lessons.
D. revising lesson plans.

_____ **8.** You have been assigned to teach a CPR class. You should gather your materials and resources during the ________________ step of the four-step instructional process.

A. preparation
B. presentation
C. application
D. evaluation

_____ **9.** The least powerful learning channel to the brain is the sense of:

A. hearing.
B. sight.
C. smell.
D. taste.

_____ **10.** The most effective way to keep adult learners informed about the requirements and expectations of a program is to:

A. provide a method of instruction.
B. meet individually with each learner.
C. provide a course syllabus.
D. review the teaching-learning process.

_____ **11.** The two primary reasons for choosing a particular instructional medium are the:

A. class size and the weather.
B. subject content and behavioral objectives.
C. class size and instructor/learner ratio.
D. subject content and location.

_____ **12.** At the third intermediate level of instruction, the learner should be able to:

A. recognize types of ladders.
B. develop prefire plans.
C. apply hydraulic formulas.
D. identify tools by name.

_____ **13.** Which of the following should a learner be capable of performing at the advanced (fifth) level of instruction?

A. Identifying tools by name
B. Locating equipment on apparatus
C. Using ground ladders properly
D. Developing prefire plans

_____ **14.** The purpose of the presentation step of the instructional process is to:

A. prepare the mind of the learner.
B. involve learners in the learning process.
C. present new skills concepts.
D. evaluate teaching.

_____ **15.** Cognitive objectives emphasize:

A. acting.
B. feeling.
C. thinking.
D. sensing.

_____ **16.** In which learning domain would a statement such as, "A fire fighter is willing to make in-service inspections," occur?

A. Cognitive
B. Affective
C. Psychomotor
D. Terminal

_____ **17.** Which of the following statements **is not** true regarding the functions of lesson plans?

A. Lesson plans serve a useful purpose in preparing substitute instructors.
B. Lesson plans promote uniformity in all courses taught.
C. Lesson plans act as a guide to instructors for delivery and evaluations.
D. Lesson plans are a rigid procedure that must be followed for successful teaching.

_____ **18.** An instructor must also have subject matter, knowledge, and:

A. the ability to communicate effectively.
B. specialized courses leading to the issuance of a teaching certificate.
C. expertise in all specialty areas.
D. All of the above

_____ **19.** The reference section of a lesson plan should include:

A. references listed by the program developers.
B. resources the instructor used to research the lesson.
C. jurisdictional protocols and operational procedures.
D. All of the above

_____ **20.** The terms "recognize," "identify," and "label" correspond to which level of instruction?

A. Level One (Basic)
B. Level Two (Basic)
C. Level Three (Intermediate)
D. Level Four (Intermediate)

_____ **21.** A positive attribute of training is a learner/instructor relationship built on:

A. mutual respect and confidence.
B. enthusiasm and personal experience.
C. impressions and attitudes.
D. ingenuity and creativity.

_____ **22.** The part of an objective that describes the important aspects of the work environment, such as what equipment and assistance will be provided to the learner, the standard of performance required, and whether notes or textbooks can be used, is known as the:

A. standard.
B. behavior.
C. condition.
D. component.

_____ **23.** At which level of instruction or knowledge would the following terms most likely be found: "recall," "select," "recognize," "list," "identify"?

A. Level One (Basic)
B. Level Two (Basic)
C. Level Three (Intermediate)
D. Level Four (Intermediate)

_____ **24.** Which of the following **is not** one of the five steps in the Five Step Planning Model?

A. Identify training needs
B. Select performance objectives
C. Implement strategic plan
D. Evaluate program

_____ **25.** Resources often overlooked by instructors that can be used to reveal safety problems in need of corrective action are:

A. records of previous accident investigations.
B. weak behavioral objectives.
C. changes in the professional qualification standards.
D. the number of emergency responses made during a given time.

_____ **26.** Students have the opportunity to have hands-on training during the _______________ step of the four-step instructional process.

A. preparation
B. presentation
C. application
D. evaluation

_____ **27.** Which of the following is a good seating arrangement when instructing a class composed of a small discussion group?

A. A chevron arrangement
B. Small rows to prevent eye contact
C. Group clusters or a U-shaped pattern
D. Existing seating since changes detract from instruction

_____ **28.** A fire service instructor should:

A. inform the learners of the concept of safety.
B. ensure the learners properly use personal protective equipment.
C. stress safety, and ensure that it is the highest priority.
D. All of the above

_____ **29.** Safety is the highest priority when training; it is imperative that what is taught be:

A. consistent with the written departmental safety policy.
B. approved by FEMA standards.
C. practiced by the student until it becomes second nature.
D. emphasized with bold type in the lesson plan if it is a safety concern.

_____ **30.** Frequently, instructors overlook the importance of _______________ when they conduct class sessions in recreation rooms, apparatus rooms, or dormitories.

A. policies
B. group control
C. suitable tables and chairs
D. session planning

_____ **31.** Group involvement is <u>**best**</u> accomplished by using the ____________ method.

A. guided discussion
B. simulation
C. case study
D. role playing

_____ **32.** Which of the following **would not** be considered an area of learner frustration?

A. Fear or worry
B. Positive feedback
C. Poor instruction
D. Discomfort of the physical environment

_____ **33.** The value of humor in the classroom is to:

A. identify with the students.
B. allow for lack of knowledge by the instructor.
C. make learning more interesting.
D. satisfy Maslow's theory of basic human needs.

_____ **34.** Instructors are expected to know the material being taught and should:

A. be judged in terms of their credentials and number of years of education.
B. possess the ability to communicate knowledge and skills to others.
C. not be required to prove competence.
D. have a minimum number of years of fire service experience.

_____ **35.** The basic instructional method to use in teaching a new task is:

A. simulating.
B. role playing.
C. lecturing.
D. demonstrating.

_____ **36.** Repetition is basic to the development of adequate responses **best** describes Thorndike's Law of:

A. Readiness.
B. Exercise.
C. Effect.
D. Practice.

_____ **37.** Based on Maslow's Hierarchy of Needs, recognizing and praising a student in front of peers can:

A. accommodate social needs.
B. provide self-actualization.
C. satisfy physiological needs.
D. build self-esteem.

_____ **38.** Approximately 83 percent of learning is a result of:

A. touching.
B. smelling.
C. hearing.
D. seeing.

_____ **39.** The concept that learning is easier if a reward is attached illustrates the Law of:

A. Effect.
B. Exercise.
C. Readiness.
D. Association.

_____ **40.** Which of Maslow's needs are satisfied when fire fighters believe they have reached their full potential?

A. Social
B. Security
C. Self-actualization
D. Self-esteem

_____ **41.** Questioning techniques are considered important instructor tools. What type of question is asked of just one person?

A. Overhead
B. Rhetorical
C. Direct
D. Redirected

_____ **42.** In the cognitive domain, there are _______________ levels of learning that can be measured through testing.

A. three
B. five
C. six
D. four

_____ **43.** A change in behavior that occurs as a result of acquiring new information and putting it to use through practice is a definition for:

A. motivation.
B. Law of Exercise.
C. learning.
D. Law of Effect.

_____ **44.** The standard that provides guidance and direction for safety officers, supervisors, and lead instructors is:

A. NFPA 1041.
B. NFPA 1500.
C. NFPA 1561.
D. NFPA 1021.

_____ **45.** Which of the following is considered to be an attribute of an effective instructor?

A. Sense of humor
B. Empathy
C. Consistency in instructional technique
D. All of the above

_____ **46.** When a student asks a question in class that refers to material which will be covered in detail later in the course, the instructor should:

1. announce that it will be covered later.
2. answer the question briefly.
3. make a note to refer to the question when presenting the material.

A. Only statement 1 is correct.
B. Only statement 2 is correct.
C. Only statements 1 and 3 are correct.
D. All three statements are correct.

_____ **47.** The best way to handle a failure of audio-visual equipment is to:

A. move to the next course section.
B. call the training officer.
C. implement a contingency plan.
D. reschedule the class.

_____ **48.** Coaching will include:

A. mastery of the subject being taught.
B. critical phrases.
C. being a follower rather than a leader.
D. keeping a distance from the learner.

_____ **49.** Which of the following verbs describes an action which the instructor must take to create student desire for learning?

A. Order
B. Motivate
C. Direct
D. Assign

_____ **50.** A successful technique that has been used to get daydreamers back "on track" is to:

A. ask a direct question.
B. remind them of their learning responsibility.
C. give frequent praise.
D. ask rhetorical questions.

_____ **51.** A learner who uses a group situation to gain attention for him/herself is classified as:

A. a staller.
B. timid.
C. a troublemaker.
D. a show-off.

_____ **52.** The uninterested learner displays little _______________ or _______________ and has a low learning and accomplishment rate.

A. enthusiasm; I.Q.
B. common sense; good judgment
C. energy; attention
D. ability; desire

_____ **53.** If a skill is being taught, make sure the students have sufficient time:

A. to take notes.
B. to practice.
C. for breaks.
D. for verbal emphasis.

_____ **54.** Instructors should slow their speaking speed when:

A. beginning a class.
B. emphasizing important items.
C. students are taking notes.
D. All of the above

_____ **55.** One of the **disadvantages** of projected instructional media is that:

A. they are visible to a large audience.
B. images are vivid.
C. they stimulate multiple senses simultaneously.
D. they are an expensive investment.

_____ **56.** For new instructors to become proficient at using multiple types of media, they should:

A. practice using the media.
B. experiment and have fun with them.
C. not use the media if they don't know how to use it.
D. Both A and B are correct

_____ **57.** While you are displaying and explaining a wall chart, you realize that the learners are being distracted by the overhead projector you have left on after the last point you made. The correct procedure would be to:

A. move your body between the projector and the screen.
B. ask the students to pay attention to the wall chart rather than the overhead image.
C. walk over to the projector and turn it off.
D. call a short break.

_____ **58.** _______________ are convenient teaching aids because they provide a permanent record and are portable.

A. Chalk board
B. Dry erase boards
C. Electronic copy boards
D. Easel pads

_____ **59.** Training prerequisites are often used to determine participants for:

A. entry into training programs.
B. eligibility for hiring.
C. promotion.
D. All of the above

_____ **60.** A test that measures what it is supposed to measure is a _______________ test.

A. comprehensive
B. discriminating
C. valid
D. reliable

_____ **61.** Incorrect answer choices to multiple-choice test questions are called:

A. distractors.
B. keyed responses.
C. detractors.
D. alternatives

_____ **62.** _________________ tests may be used to supplement performance tests.

A. Oral
B. Written
C. Essay
D. Multiple choice

_____ **63.** Which of the following <u>is</u> <u>not</u> a purpose of written tests?

A. To find out whether students can perform the job taught
B. To provide the only means of evaluation
C. To identify student weaknesses
D. To hold learners accountable for material taught

_____ **64.** Most cheating on tests is the result of:

A. copying during test administration.
B. hints given by test administrators.
C. advance knowledge of test content.
D. Both A and C are correct

_____ **65.** A pretest is used by instructors to:

A. determine the current level of training and skills.
B. make decisions on program content.
C. specify a degree of satisfactory accomplishment.
D. Both A and B are correct

_____ **66.** _______ score consists of the points a learner receives on a test.

A. Percentage
B. Raw
C. Average
D. Mean

_____ **67.** The Family Educational Rights and Privacy Act of 1974 requires that the release of any information from a learner's record is allowable only:

A. when authorized by a municipality's legal advisor.
B. to the highest elected official in a municipality.
C. with permission from the learner.
D. with written permission from a family member.

_____ **68.** Which of the following would be an acceptable method for releasing test scores from a test that you administered and graded?

A. Near the classroom door, post the scores next to the students' social security numbers.
B. Read the names and scores to the class as a group.
C. Announce the range of the scores to the class as a whole, but give each student his/her score privately.
D. All of the above

_____ **69.** Formative evaluation measures learner performance in an effort to answer which of the following questions?

A. Are the learners competent?
B. Have the learners learned in the most efficient way?
C. Are the learners frustrated with the presentation?
D. All of the above

_____ **70.** When is a summative evaluation process employed?

A. Before the course
B. Periodically during course development
C. After the course
D. Continuously

_____ **71.** If a student answers 85 out of 100 test items correctly, the **raw score** is:

A. 85.
B. 100.
C. 15.
D. 100/85.

_____ **72.** Job performance requirements:

A. are required for a specific job, based on standards.
B. are part of the session scheduling.
C. will involve the discussion process.
D. are prone to involve the halo process.

_____ **73.** Tests should be based upon:

A. text materials.
B. lesson plans.
C. behavioral objectives.
D. principles of instruction.

_____ **74.** A type of evaluation that can be used throughout a course delivery and provides immediate feedback is a ________________ evaluation.

A. comprehensive
B. formative
C. prescriptive
D. summative

_____ **75.** An instructor can gauge how thoroughly the learners are grasping the lesson by:

A. the success/fail rate of the learners.
B. the feedback received from learners.
C. evaluations by other instructors.
D. Both A and B are correct

Now that you have finished the feedback step for Examination I-1, it is time to repeat the process by taking another comprehensive examination for NFPA Standard 1041.

Did you score higher than 80% on Examination I-1? Circle Yes or No in ink. (We will return to your answer to this question later in SAEP).

Examination I-2, Adding Difficulty and Depth

During Examination I-2, progress will be made in developing depth of knowledge and skills.

Step 1—Take Examination I-2. When you have completed Examination I-2, go to Appendix A and compare your answers with the correct answers.

Step 2—Score Examination I-2. How many examination items did you miss? Write the number of missed examination items in the blank in ink. ________________ Enter the number of examination items you guessed in this blank. ________________ Enter these numbers in the designated locations on your Personal Progress Plotter.

Step 3—Once again the learning begins. During the feedback step, research the correct answer using Appendix A information for Examination I-2. Highlight the correct answer during your research of the reference materials. Read the entire paragraph containing the correct answer.

Helpful Hint

Follow each step carefully to realize the best return on effort. Would you consider investing your money in a venture without some chance of return on that investment? Examination preparation is no different. You are investing time and expecting a significant return for that time. If, indeed, time is money, then you are investing money and are due a return on that investment. Doing things right and doing the right things in examination preparation will ensure the maximum return on effort.

Examination I-2

Directions

Remove Examination I-2 from the manual. First, take a careful look at the examination. There should be 75 examination items. Notice that a blank line precedes each examination item number. This line is provided for you to enter the answer to the examination item. Write the answer in ink. Remember the rule about not changing your answers. Our research has shown that changed answers are often incorrect, and, more often than not, the answer that is chosen first is correct.

If you guess the answer to a question, place an "X" or a check mark by your answer. This step is vitally important as you gain and master knowledge. We will explain how we treat the "guessed" items later in SAEP.

Take the examination. Once you complete it, go to Appendix A and score your examination. Once the examination is scored, carefully follow the directions for feedback on the missed and guessed examination items.

_____ **1.** A goal is:

A. a broad statement of educational intent.
B. the intended end result of instruction.
C. a standard or degree of performance.
D. a given situation under which performance is to occur.

_____ **2.** Directions: Read the following statements regarding the teaching of skills, then choose the appropriate answer from A–D below.

1. An instructor should present the skill at a mastery level.
2. An instructor may introduce a skill in steps.
3. Substitute instructors could make the class a discovery zone, using each objective as an overview point.

A. All three statements are true.
B. Statement 1 is true; statements 2 and 3 are false.
C. Statement 2 is true; statements 1 and 3 are false.
D. Statement 3 is true; statements 1 and 2 are false.

_____ **3.** Which NFPA Standard is specifically designed for the safety officer?

A. NFPA 1521
B. NFPA 1500
C. NFPA 1561
D. NFPA 1506

_____ **4.** Training records can be maintained on a computer system:

A. if security measures are in place to limit access.
B. only if allowed by local authority.
C. if access is regulated by policy.
D. Both A and C are correct

_____ **5.** An accident investigator should include identification of all of the following **except**:

A. the principal sources of accidents.
B. who was at fault for the accident.
C. problems in operating procedures.
D. unsafe conditions of equipment.

_____ **6.** Analyzing circumstances surrounding accidents can enable an instructor to:

A. identify and locate principle sources of accidents.
B. disclose the nature and size of accident problems in different operations.
C. indicate the need for engineering revisions by identifying unsafe conditions of various types of equipment.
D. All of the above

_____ **7.** The instructional sequencing method that starts with an overview of a topic, then discusses each individual topic and finishes with a review of the main point is called:

A. whole-part-whole.
B. simple-to-complex.
C. cognitive-to-psychomotor.
D. known-to-unknown.

_____ **8.** People learn more effectively by:

A. listening to lectures.
B. passively watching videos.
C. reading books.
D. interacting with other people.

_____ **9.** The role of an instructor who provides advanced-level training for experienced fire and emergency service responders is **best** described as:

A. facilitator.
B. teacher.
C. safety officer.
D. technical advisor.

_____ **10.** In which step of the lesson plan outline are learners motivated to learn?

A. Preparation
B. Presentation
C. Application
D. Evaluation

_____ **11.** During which step in a lesson plan is a learner given an opportunity to practice what has been learned?

A. Preparation
B. Presentation
C. Application
D. Evaluation

____ **12.** Under which of the following premises is individualized instruction developed?

A. Learner's needs, learning style, learning objectives, and competencies can be accommodated.
B. Organizational needs, lesson plan format, and instructor skills can be served or utilized.
C. Learning styles and learner's preferences are most important.
D. Learner's preferences and instructor skills can be accommodated.

____ **13.** What is the principal difference between a study sheet ("S") and an information sheet ("I")?

A. "S" is more detailed than "I."
B. "S" is used by students, "I" by instructors.
C. "I" is a fact sheet to supplement other course resources; "S" is designed to arouse learner interest in areas to study.
D. "S" provides complete information, while "I" tells where to find the information.

____ **14.** To be effective, objectives should be stated:

A. as simply as possible.
B. at the beginning of a test.
C. in terms of measurable, observable performance.
D. to meet general guidelines.

____ **15.** **Directions:** Read the following statements regarding instructors, then choose the appropriate answer from A–D below.

1. Instructors lead or guide learners through a class.
2. Instructors must have a desire to understand the needs of learners.
3. An instructor needs special training that leads to a teaching certificate.

A. All three statements are true.
B. Statement 1 is true; statements 2 and 3 are false.
C. Statements 1 and 2 are true; statement 3 is false.
D. Statement 3 is true; statements 1 and 2 are false.

____ **16.** A lesson plan format includes all of the following **except**:

A. a list of materials needed.
B. references.
C. evaluations.
D. learner evaluations of instructor.

____ **17.** The domain of learning which emphasizes physical skills is the ________________ domain.

A. cognitive
B. affective
C. psychomotor
D. enabling

_____ **18.** Condition, behavior (performance), and standard (criteria) are three parts of the:

A. communications theory.
B. learning process.
C. objective.
D. motivation theory.

_____ **19.** A lesson plan makes effective use of:

A. time, personnel, and space.
B. facilities, finances, and environment.
C. materials, lighting, and equipment.
D. All of the above

_____ **20.** A lesson plan is:

A. a guide showing the way learners and instructors will spend their class time.
B. an outline of material and procedures that the instructor plans to follow.
C. a guide to uniformity in teaching.
D. All of the above

_____ **21.** Which one of the following provides the learner with a step-by-step procedure for learning a manipulative or psychomotor skill?

A. Job breakdown sheet
B. Competency profile
C. Study sheet
D. Study objectives

_____ **22.** All of the items below are major elements that guide instructors in preparing to teach **except** session:

A. scheduling.
B. analysis.
C. preparation.
D. logistics.

_____ **23.** Which of the following **would not** usually be given consideration during the session preparation phase?

A. Reading the lesson objectives
B. Checking on equipment needed
C. Scheduling classroom and facilities
D. Determining skills to be taught

_____ **24.** A key factor in the classroom climate is:

A. discipline.
B. theatrics.
C. procedures.
D. enthusiasm.

_____ **25.** When preparing to teach, one of an instructor's **primary** responsibilities is to provide the learner with:

A. handouts.
B. comfortable seating.
C. an appropriate physical setting.
D. media-based instruction.

_____ **26.** The fire service instructor has responsibilities to the fire service, the organization, and to the learners. Which one of the following concerns is **primary** to all three groups listed above?

A. Leadership skills
B. Planning training programs
C. Safety
D. Knowledge of fire service skills

_____ **27.** The NFPA Standard concerning fire department occupational safety is:

A. NFPA 1021.
B. NFPA 1521.
C. NFPA 1403.
D. NFPA 1500.

_____ **28.** "Bullying" by an instructor is:

A. needed to earn the respect of unruly students.
B. an expression of fear or frustration.
C. an indication of lack of knowledge.
D. practiced only as a last resort.

_____ **29.** The **primary** consideration when performing a demonstration is to demonstrate the ________________ way.

A. safest
B. easiest
C. quickest
D. most accurate

_____ **30.** If boredom appears prevalent during a class, the instructor should:

A. speak louder and more authoritatively.
B. call for a short break.
C. use humor to alleviate boredom.
D. explain the necessity of paying attention.

_____ **31.** When weather conditions are inclement, the instructor may assess the available options by:

A. providing frequent breaks.
B. moving the group inside.
C. taking other steps to provide comfort.
D. All of the above

_____ **32.** Which of the following **is not** part of the Evaluation step of the instructional process?

A. Having the learner perform the job unassisted
B. Asking prepared questions
C. Explaining procedures
D. Having learners criticize other learners' performances

_____ **33.** Maslow's "Hierarchy of Needs" includes:

A. compensation, social, and psychological.
B. security, self-esteem, and self-actualization.
C. valuing, responding, and self-actualization.
D. self-esteem, money, and recognition.

_____ **34.** A learner/instructor relationship based upon mutual respect, confidence, and rapport is:

A. not as important if the learners outrank the instructor.
B. not as important if the instructor outranks the learners.
C. if the instructor is thoroughly qualified.
D. important as a positive attribute for learning.

_____ **35.** Learning is **best** accomplished through ________________ sessions.

A. short, intensive
B. long, extensive
C. short, extensive
D. long, intensive

_____ **36.** The communications process exists between two people who are known as the ________________ and the ________________.

A. instructor; receiver
B. sender; receiver
C. developer; interpreter
D. speaker; interpreter

_____ **37.** The Law of Readiness is based upon a principle of learning which states that people learn best when:

A. they are prepared to learn.
B. they have the proper environment for learning.
C. the instructor is ready to teach.
D. the need for the learning is adequately explained.

_____ **38.** Learning will always be effective when a feeling of satisfaction, pleasantness, or reward accompanies the learning process. This statement describes the:

A. Law of Learning.
B. Law of Readiness.
C. Law of Exercise.
D. Law of Effect.

_____ **39.** An active process that produces a change in behavior as a result of acquiring new information **best** describes:

A. teaching.
B. understanding.
C. learning.
D. Pavlov's experiment.

_____ **40.** Abraham Maslow has identified basic human needs. These needs are referred to as the ________________ of Needs.

A. Triangle
B. Pyramid
C. Hierarchy
D. Ladder

_____ **41.** Motivation is created:

A. by the instructor.
B. from within the student.
C. with a lesson plan.
D. through the use of training materials.

_____ **42.** According to Maslow's Hierarchy of Needs, the lowest or basic need in a human is:

A. security.
B. physiological.
C. social.
D. self-esteem.

_____ **43.** During the communication process, messages that are transmitted without words (gestures, eye contact, and tone of voice) are called:

A. emotions.
B. nonverbal cues.
C. semantics.
D. attitudes.

_____ **44.** For instructors to be effective, they must understand the meaning of words, how to use them, and in what context a word should be used. This is a definition of:

A. nonverbal communication.
B. the study of emotional communication.
C. course objective verbalization.
D. semantics.

_____ **45.** In extreme environmental conditions, the instructor should:

A. show as much of the skill as possible under some protection in an outside area.
B. provide participants with frequent rest breaks.
C. consider moving the class to another location.
D. All of the above

_____ **46.** The _______________ is responsible for maintaining continuity throughout the training program.

A. program developer
B. company officer
C. ranking officer
D. instructor

_____ **47.** A fire service instructor is expected to meet the needs of all of the following **except**:

A. the organization.
B. their personal acceptance.
C. the learner.
D. management.

_____ **48.** Extroverted individuals can be managed by:

A. reminding individuals of the lesson topic.
B. embarrassing the individual when he or she distracts the class.
C. offering counseling.
D. encouraging individual participation.

_____ **49.** An adult learner becomes motivated when he or she:

A. is bored.
B. can identify by "what's in it for me."
C. has to be there.
D. notices other students participating.

_____ **50.** It is not advisable to permit gifted learners to:

A. work ahead of the rest of the class.
B. handle challenging assignments.
C. accomplish more than is expected of average students.
D. spend time on assignments that are below their ability.

_____ **51.** A shy or timid learner can be helped by:

A. coupling with a fast learner.
B. a personal talk.
C. encouraging participation when discussion is informal.
D. giving frequent praise.

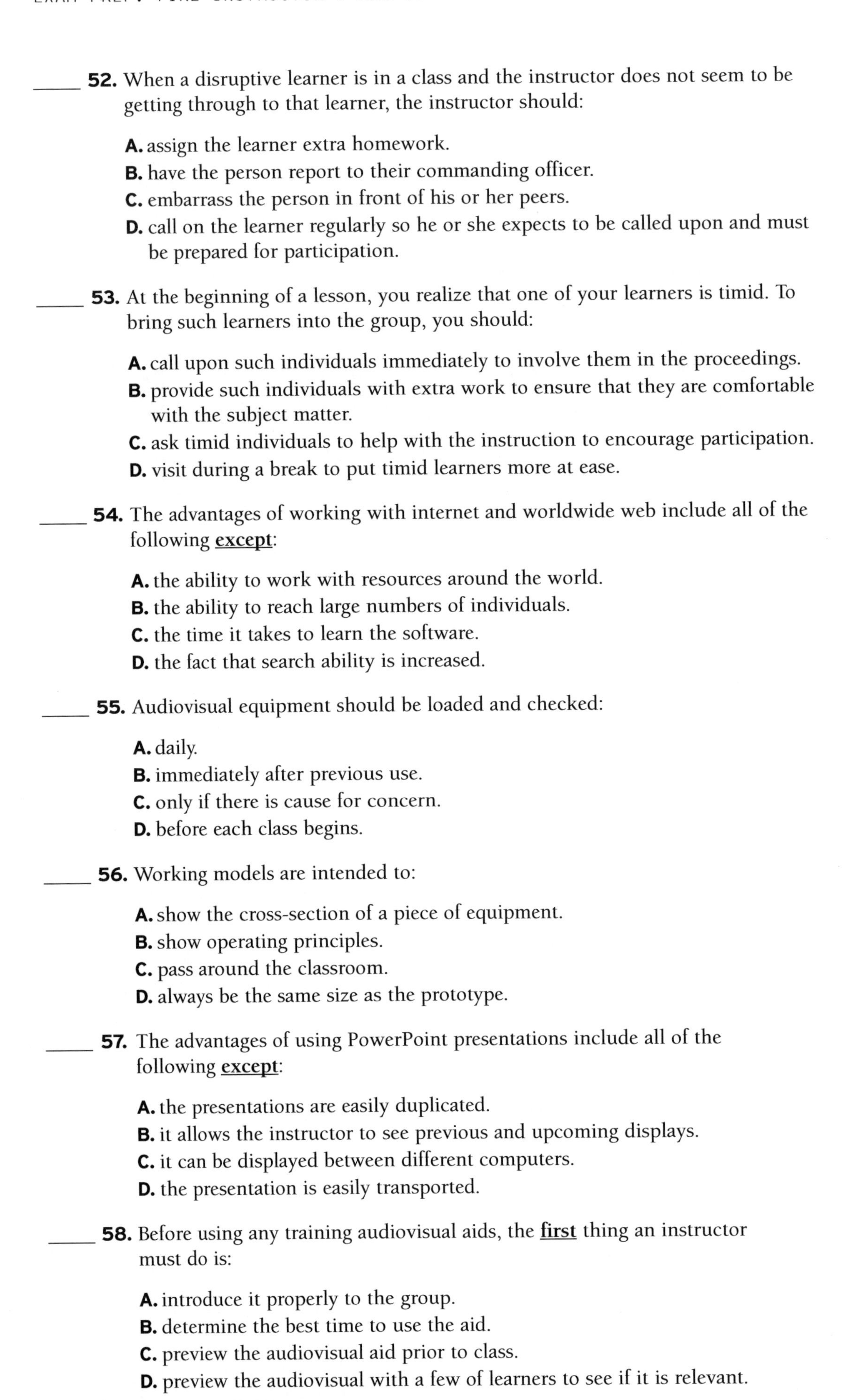

____ **52.** When a disruptive learner is in a class and the instructor does not seem to be getting through to that learner, the instructor should:

A. assign the learner extra homework.
B. have the person report to their commanding officer.
C. embarrass the person in front of his or her peers.
D. call on the learner regularly so he or she expects to be called upon and must be prepared for participation.

____ **53.** At the beginning of a lesson, you realize that one of your learners is timid. To bring such learners into the group, you should:

A. call upon such individuals immediately to involve them in the proceedings.
B. provide such individuals with extra work to ensure that they are comfortable with the subject matter.
C. ask timid individuals to help with the instruction to encourage participation.
D. visit during a break to put timid learners more at ease.

____ **54.** The advantages of working with internet and worldwide web include all of the following **except**:

A. the ability to work with resources around the world.
B. the ability to reach large numbers of individuals.
C. the time it takes to learn the software.
D. the fact that search ability is increased.

____ **55.** Audiovisual equipment should be loaded and checked:

A. daily.
B. immediately after previous use.
C. only if there is cause for concern.
D. before each class begins.

____ **56.** Working models are intended to:

A. show the cross-section of a piece of equipment.
B. show operating principles.
C. pass around the classroom.
D. always be the same size as the prototype.

____ **57.** The advantages of using PowerPoint presentations include all of the following **except**:

A. the presentations are easily duplicated.
B. it allows the instructor to see previous and upcoming displays.
C. it can be displayed between different computers.
D. the presentation is easily transported.

____ **58.** Before using any training audiovisual aids, the **first** thing an instructor must do is:

A. introduce it properly to the group.
B. determine the best time to use the aid.
C. preview the audiovisual aid prior to class.
D. preview the audiovisual with a few of learners to see if it is relevant.

_____ **59.** The _________________ overhead projector is designed to be used in a lighted room.

A. incandescent
B. mercury
C. halogen
D. florescent

_____ **60.** One **disadvantage** of a videotape system is:

A. losing learners' attention.
B. that segments of videotape may be replayed.
C. depicting material with authenticity.
D. that it requires extensive time to script, shoot, and edit.

_____ **61.** For a film or videotape to be an effective learning tool, the instructor must:

A. show the film or videotape at least twice.
B. view the film or videotape before the class begins.
C. prepare the learners for what they are about to see.
D. Both B and C are correct

_____ **62.** A problem encountered with using small "pass-around" items as instructional media is learners:

A. talk among themselves about nonsubject material.
B. take the item(s) resulting inventory loss.
C. have their attention divided between the instructor and the item(s) they are examining.
D. drop the item(s) and possibly cause damage.

_____ **63.** Which of the following is (are) typical of experienced fire service responders during training exercises?

A. They are cooperative learners.
B. They are task oriented.
C. They are mastering individual skills.
D. Both A and B are correct

_____ **64.** Cheating on tests can be minimized by:

A. regularly revising tests.
B. special seating arrangements.
C. careful monitoring during a test.
D. All of the above

_____ **65.** To aid learners in reaching their highest potential in taking a test, the instructor should do all of the following **except**:

A. read instructions aloud while students read them silently.
B. provide directions that are clear.
C. give hints about individual items when answering questions.
D. keep instructions to a minimum during the test.

_____ **66.** A formative evaluation will answer the question:

A. Have the participants learned in the most efficient way?
B. Have the participants learned what they need to know to do their work?
C. Did the test measure what it was meant to measure?
D. Do the test results measure the degree of learning?

_____ **67.** A written test that may be difficult to score because more than one answer may be correct is the:

A. short answer/completion test.
B. true/false test.
C. matching test.
D. multiple choice test.

_____ **68.** The method of evaluation used to determine a person's ability to accomplish a task is the ________________ test.

A. subjective
B. objective
C. performance
D. written

_____ **69.** The Family Educational Rights and Privacy Act of 1974 requires that the release of any information from a learner's record is allowable only:

A. when authorized by a municipality's legal advisor.
B. to the highest elected official in a municipality.
C. with permission from the learner.
D. with written permission from a family member.

_____ **70.** Coaching includes all of the following **except**:

A. observation.
B. evaluation.
C. providing written directions.
D. suggestions for improvement.

_____ **71.** Which of the following qualities is recommended for an effective question?

A. Leads to a certain response
B. Requires only a "yes" or "no" answer
C. Should not make learners feel ignorant
D. Does not solicit opinions abut the topic being discussed

_____ **72.** Under the Family Educational and Privacy Act of 1974 an individual's name and ________________ cannot be displayed **together**.

A. attendance
B. grades
C. age
D. marital status

_____ **73.** If a class scores poorly on a test, the instructor should:

A. throw out the poor test items (questions).
B. reteach the lesson.
C. change the distractors and give the test again.
D. revise the lesson plan.

_____ **74.** The evaluation step of a lesson plan:

A. tests the effectiveness of the learning environment.
B. determines whether the objectives of the lesson have been achieved.
C. reveals whether the students can do the job unaided and without supervision.
D. provides for learner evaluation of the instructor.

_____ **75.** Which of the following is not considered a primary means of helping learners with learning disabilities?

A. Providing feedback on progress
B. Teaching to the slowest learning level
C. Individualized instruction
D. Tutoring

Helpful Hint

Try to determine why you selected the wrong answer. Usually something influenced your selection. Focus on the difference between your wrong answer and the right answer. Carefully read and study the entire paragraph containing the correct answer. Highlight the answer just as you did for Examination I-1.

Did you score higher than 80% on Examination I-2 Feedback? Circle Yes or No in ink. (We will return to your answer to this question later in SAEP.)

Examination I-3, Confirming What You Mastered

During Examination I-3, progress will be made in reinforcing what you have learned and improving your examination-taking skills. This examination contains approximately 60 percent of the examination items you have already answered and several new examination items. Follow the steps carefully to realize the best return on effort.

Step 1—Take Examination I-3. When you have completed Examination I-3, go to Appendix A and compare your answers with the correct answers.

Step 2—Score Examination I-3. How many examination items did you miss? Write the number of missed examination items in the blank in ink. ________________ Enter the number of examination items you guessed in this blank. ________________ Enter these numbers in the designated locations on your Personal Progress Plotter.

Step 3—During the feedback step, research the correct answer using the Appendix A information for Examination I-3. Highlight the correct answer during your research of the reference materials. Read the entire paragraph containing the correct answer.

Examination I-3

Directions

Remove Examination I-3 from the manual. First, take a careful look at the examination. There should be 100 examination items. Notice that a blank line precedes each examination item number. This line is provided for you to enter the answer to the examination item. Write the answer in ink. Remember the rule about not changing your answers. Our research has shown that changed answers are often incorrect, and, more often than not, the answer that is chosen first is correct.

If you guess the answer to a question, place an "X" or a check mark by your answer. This step is vitally important as you gain and master knowledge. We will explain how we treat the "guessed" items later in SAEP.

Take the examination. Once you complete it, go to Appendix A and score your examination. Once the examination is scored, carefully follow the directions for feedback on the missed and guessed examination items.

_____ **1.** Performance standards for fire instructors are identified in:

A. NFPA 1021.
B. NFPA 1041.
C. NFPA 1403.
D. NFPA 1031.

_____ **2.** The **primary** role of the instructor is:

A. planning and conducting training.
B. pursuing knowledge to adapt changes in life and work.
C. reviewing changes in regional curricula.
D. acting as a mentor to aspiring instructors.

_____ **3.** A goal is:

A. a broad statement of educational intent.
B. the intended end result of instruction.
C. a standard or degree of performance.
D. a given situation under which performance is to occur.

_____ **4.** **Directions:** Read the following statements regarding the teaching of skills, then choose the appropriate answer from A–D below:

1. An instructor should present the skill at a mastery level.
2. An instructor may introduce a skill in steps.
3. Substitute instructors could make the class a discovery zone, using each objective as an overview point.

A. All three statements are true.
B. Statement 1 is true; statements 2 and 3 are false.
C. Statement 2 is true; statements 1 and 3 are false.
D. Statement 3 is true; statements 1 and 2 are false.

_____ **5.** Which NFPA Standard is specifically designed for the safety officer?

A. NFPA 1521
B. NFPA 1500
C. NFPA 1561
D. NFPA 1506

_____ **6.** The **primary** purpose of collecting training data and analyzing the information is to:

A. adjust evaluation techniques.
B. meet State Fire Marshal requirements.
C. enhance learning efficiency.
D. provide information to NFIRS.

_____ **7.** NFPA 1403 gives information for establishing procedures for:

A. live fire exercises.
B. wildland firefighting.
C. water supply and firefighting in rural areas.
D. the incident management system.

_____ **8.** Training records can be maintained on a computer system:

A. if security measures are in place to limit access.
B. only if allowed by local authority.
C. if access is regulated by policy.
D. Both A and C are correct

_____ **9.** An accident analysis is conducted to achieve all of the following **except**:

A. disclose the nature and size of the accident problem.
B. determine lack of action.
C. identify the need for engineering revisions.
D. identify problems in operating procedures/guidelines.

_____ **10.** An accident investigator should include identification of all of the following **except**:

A. the principal sources of accidents.
B. who was at fault for the accident.
C. problems in operating procedures.
D. unsafe conditions of equipment.

_____ **11.** You have been assigned to teach a CPR class. You should gather your materials and resources during the _______________ step of the four-step instructional process.

A. preparation
B. presentation
C. application
D. evaluation

_____ **12.** The <u>least</u> powerful learning channel to the brain is the sense of:

A. hearing.
B. sight.
C. smell.
D. taste.

_____ **13.** The two <u>primary</u> reasons for choosing a particular instructional medium are the:

A. class size and the weather.
B. subject content and behavioral objectives.
C. class size and instructor/learner ratio.
D. subject content and location.

_____ **14.** At the third intermediate level of instruction, the learner should be able to:

A. recognize types of ladders.
B. develop prefire plans.
C. apply hydraulic formulas.
D. identify tools by name.

_____ **15.** During which step in a lesson plan is a learner given an opportunity to practice what has been learned?

A. Preparation
B. Presentation
C. Application
D. Evaluation

_____ **16.** Whether the training is academic or practical, success during the presentation step requires thorough planning and should:

A. follow a logical sequence.
B. employ only the lecture mode.
C. be delivered in a classroom.
D. always proceed from the known to the unknown.

_____ **17.** The purpose of the presentation step of the instructional process is to:

A. prepare the mind of the learner.
B. involve learners in the learning process.
C. present new skills concepts.
D. evaluate teaching.

_____ **18.** <u>Directions:</u> Read the following statements regarding instructors, then choose the appropriate answer from A–D below:

1. Instructors lead or guide learners through a class.
2. Instructors must have a desire to understand the needs of learners.
3. An instructor needs special training that leads to a teaching certificate.

A. All three statements are true.
B. Statement 1 is true; statements 2 and 3 are false.
C. Statements 1 and 2 are true; statement 3 is false.
D. Statement 3 is true; statements 1 and 2 are false.

_____ **19.** An effective instructor is a sincere person with:

A. a desire to teach.
B. in-depth knowledge of the subject.
C. a sense of empathy.
D. All of the above

_____ **20.** Cognitive objectives emphasize:

A. acting.
B. feeling.
C. thinking.
D. sensing.

_____ **21.** The domain of learning which emphasizes physical skills is the _______________ domain.

A. cognitive
B. affective
C. psychomotor
D. enabling

_____ **22.** A lesson plan makes effective use of:

A. time, personnel, and space.
B. facilities, finances, and environment.
C. materials, lighting, and equipment.
D. All of the above

_____ **23.** An instructor must also have subject matter, knowledge, and:

A. the ability to communicate effectively.
B. specialized courses leading to the issuance of a teaching certificate.
C. expertise in all specialty areas.
D. All of the above

_____ **24.** Which one of the following provides the learner with a step-by-step procedure for learning a manipulative or psychomotor skill?

A. Job breakdown sheet
B. Competency profile
C. Study sheet
D. Study objectives

_____ **25.** The reference section of a lesson plan should include:

A. references listed by the program developers.
B. resources the instructor used to research the lesson.
C. jurisdictional protocols and operational procedures.
D. All of the above

_____ **26.** The terms "recognize," "identify," and "label" correspond to which level of instruction?

A. Level One (Basic)
B. Level Two (Basic)
C. Level Three (Intermediate)
D. Level Four (Intermediate)

_____ **27.** The Law of _______________ states that no one will ever become proficient at any skill without performing the operation.

A. Exercise
B. Readiness
C. Intensity
D. Repetition

_____ **28.** What are the components of a performance objective?

A. Occupational analysis, lesson plan, evaluation
B. Task analysis, action, completion
C. Condition, task, standard
D. Describe, identify, demonstrate

_____ **29.** A positive attribute of training is a learner/instructor relationship built on:

A. mutual respect and confidence.
B. enthusiasm and personal experience.
C. impressions and attitudes.
D. ingenuity and creativity.

_____ **30.** At which level of instruction or knowledge would the following terms most likely be found: "recall," "select," "recognize," "list," "identify"?

A. Level One (Basic)
B. Level Two (Basic)
C. Level Three (Intermediate)
D. Level Four (Intermediate)

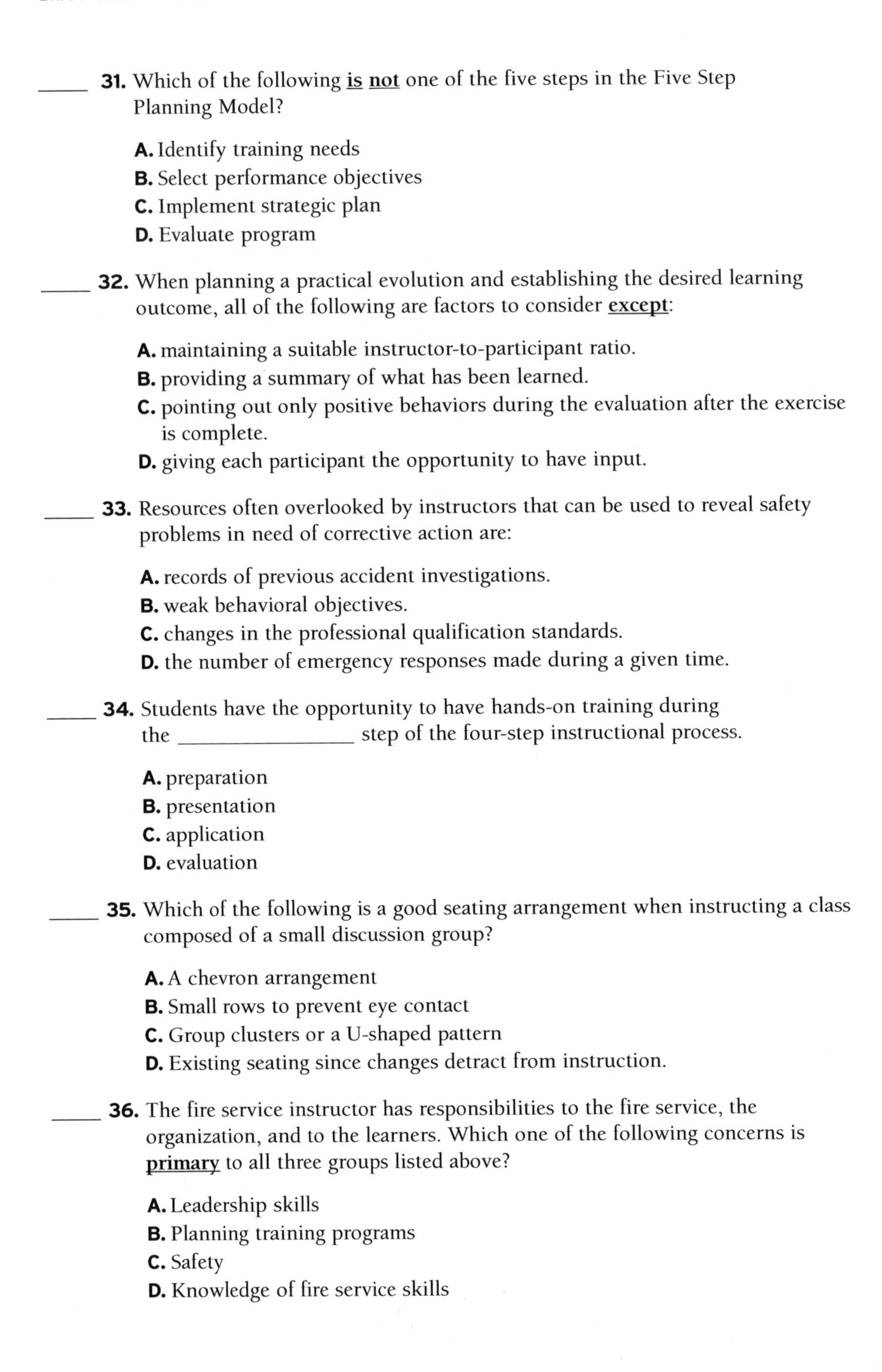

____ **31.** Which of the following **is not** one of the five steps in the Five Step Planning Model?

A. Identify training needs
B. Select performance objectives
C. Implement strategic plan
D. Evaluate program

____ **32.** When planning a practical evolution and establishing the desired learning outcome, all of the following are factors to consider **except**:

A. maintaining a suitable instructor-to-participant ratio.
B. providing a summary of what has been learned.
C. pointing out only positive behaviors during the evaluation after the exercise is complete.
D. giving each participant the opportunity to have input.

____ **33.** Resources often overlooked by instructors that can be used to reveal safety problems in need of corrective action are:

A. records of previous accident investigations.
B. weak behavioral objectives.
C. changes in the professional qualification standards.
D. the number of emergency responses made during a given time.

____ **34.** Students have the opportunity to have hands-on training during the ________________ step of the four-step instructional process.

A. preparation
B. presentation
C. application
D. evaluation

____ **35.** Which of the following is a good seating arrangement when instructing a class composed of a small discussion group?

A. A chevron arrangement
B. Small rows to prevent eye contact
C. Group clusters or a U-shaped pattern
D. Existing seating since changes detract from instruction.

____ **36.** The fire service instructor has responsibilities to the fire service, the organization, and to the learners. Which one of the following concerns is **primary** to all three groups listed above?

A. Leadership skills
B. Planning training programs
C. Safety
D. Knowledge of fire service skills

_____ **37.** A fire service instructor should:

A. inform the learners of the concept of safety.
B. ensure the learners properly use personal protective equipment.
C. stress safety, and ensure that it is the highest priority.
D. All of the above

_____ **38.** The NFPA Standard concerning fire department occupational safety is:

A. NFPA 1021.
B. NFPA 1521.
C. NFPA 1403.
D. NFPA 1500.

_____ **39.** When should a fire service instructor appraise the physical setting for a presentation?

A. As the students report for class
B. When the location leaves much to be desired
C. During the application phase
D. Before class begins

_____ **40.** An instructor must be prepared for unanticipated situations, including:

A. inclement weather.
B. learning style differences.
C. equipment variation.
D. All of the above

_____ **41.** The primary consideration when performing a demonstration is to demonstrate the _________________ way.

A. safest
B. easiest
C. quickest
D. most accurate

_____ **42.** If boredom appears prevalent during a class, the instructor should:

A. speak louder and more authoritatively.
B. call for a short break.
C. use humor to alleviate boredom.
D. explain the necessity of paying attention.

_____ **43.** A method of instruction used when students act out situations is known as:

A. simulating.
B. role playing.
C. brainstorming.
D. guided discussion.

_____ **44.** The method of instruction to use when you want learners to get practical experience without "real life" risks is:

A. simulation.
B. demonstration.
C. role playing.
D. brainstorming.

_____ **45.** When weather conditions are inclement, the instructor may assess the available options by:

A. providing frequent breaks.
B. moving the group inside.
C. taking other steps to provide comfort.
D. All of the above

_____ **46.** Which of the following **is not** part of the Evaluation step of the instructional process?

A. Having the learner perform the job unassisted
B. Asking prepared questions
C. Explaining procedures
D. Having learners criticize other learners' performances

_____ **47.** The value of humor in the classroom is to:

A. identify with the students.
B. allow for lack of knowledge by the instructor.
C. make learning more interesting.
D. satisfy Maslow's theory of basic human needs.

_____ **48.** Maslow's "Hierarchy of Needs" includes:

A. compensation, social, and psychological.
B. security, self-esteem, and self-actualization.
C. valuing, responding, and self-actualization.
D. self-esteem, money, and recognition.

_____ **49.** Learning is **best** accomplished through _______________ sessions.

A. short, intensive
B. long, extensive
C. short, extensive
D. long, intensive

_____ **50.** It is the responsibility of a fire service instructor to plan his teaching methods so that ideas are conveyed to the learner:

A. primarily through visual means.
B. through as many of the senses as practical.
C. through audio means.
D. through manipulative skills only.

_____ **51.** To introduce a subject or lesson, the instructor should state:

A. how participants will be evaluated.
B. the objective of the lesson.
C. which teaching method is about to be employed.
D. what reference material will be used.

_____ **52.** The basic instructional method to use in teaching a new task is:

A. simulating.
B. role playing.
C. lecturing.
D. demonstrating.

_____ **53.** The Law of Readiness is based upon a principle of learning which states that people learn best when:

A. they are prepared to learn.
B. they have the proper environment for learning.
C. the instructor is ready to teach.
D. the need for the learning is adequately explained.

_____ **54.** Repetition is basic to the development of adequate responses **best** describes Thorndike's Law of:

A. Readiness.
B. Exercise.
C. Effect.
D. Practice.

_____ **55.** Abraham Maslow has identified basic human needs. These needs are referred to as the _________________ of Needs.

A. Triangle
B. Pyramid
C. Hierarchy
D. Ladder

_____ **56.** Which of the following **is not** true of adult learners?

A. Adult orientation to learning is problem-centered.
B. Adult learning is very similar to adolescent learning.
C. Adults learn by relating new material to previous experience.
D. Adults have internal incentives to learn.

_____ **57.** Which of Maslow's needs are satisfied when fire fighters believe they have reached their full potential?

A. Social
B. Security
C. Self-actualization
D. Self-esteem

_____ **58.** A learner can gain more knowledge in a shorter period of time by:

A. using pure lecture as a medium of learning.
B. using hands-on training only.
C. using as many of the senses as practical.
D. studying written material.

_____ **59.** Humor has a place in the teaching/learning situation, but it should be used:

A. occasionally and in good taste.
B. when the lead instructor is not present.
C. in drill situations.
D. in counseling situations.

_____ **60.** Which of the following is considered to be an attribute of an effective instructor?

A. Sense of humor
B. Empathy
C. Consistency in instructional technique
D. All of the above

_____ **61.** In extreme environmental conditions, the instructor should:

A. show as much of the skill as possible under some protection in an outside area.
B. provide participants with frequent rest breaks.
C. consider moving the class to another location.
D. All of the above

_____ **62.** The ________________ is responsible for maintaining continuity throughout the training program.

A. program developer
B. company officer
C. ranking officer
D. instructor

_____ **63.** When a student asks a question in class that refers to material which will be covered in detail later in the course, the instructor should:

1. announce that it will be covered later.
2. answer the question briefly.
3. make a note to refer to the question when presenting the material.

A. Only statement 1 is correct.
B. Only statement 2 is correct.
C. Only statements 1 and 3 are correct.
D. All three statements are correct.

_____ **64.** Since classes may be interrupted, an instructor must be prepared to respond by:

A. having a contingency plan.
B. rescheduling activities.
C. showing a video.
D. All of the above

_____ **65.** The <u>best</u> way to handle a failure of audio-visual equipment is to:

A. move to the next course section.
B. call the training officer.
C. implement a contingency plan.
D. reschedule the class.

_____ **66.** Coaching will include:

A. mastery of the subject being taught.
B. critical phrases.
C. being a follower rather than a leader.
D. keeping a distance from the learner.

_____ **67.** Which of the following verbs describes an action which the instructor must take to create student desire for learning?

A. Order
B. Motivate
C. Direct
D. Assign

_____ **68.** Extroverted individuals can be managed by:

A. reminding individuals of the lesson topic.
B. embarrassing the individual when he or she distracts the class.
C. offering counseling.
D. encouraging individual participation.

_____ **69.** An adult learner becomes motivated when he or she:

A. is bored.
B. can identify by "what's in it for me."
C. has to be there.
D. notices other students participating.

_____ **70.** The instructor is constantly being observed by students and the impression made by the instructor will:

A. have little effect on the outcome of the class.
B. dominate the classroom atmosphere.
C. depend upon how the instructor dresses.
D. affect student response and learning initiative.

_____ **71.** It <u>is</u> <u>not</u> advisable to permit gifted learners to:

A. work ahead of the rest of the class.
B. handle challenging assignments.
C. accomplish more than is expected of average students.
D. spend time on assignments that are below their ability.

_____ **72.** When a disruptive learner is in a class and the instructor does not seem to be getting through to that learner, the instructor should:

A. assign the learner extra homework.
B. have the person report to their commanding officer.
C. embarrass the person in front of his or her peers.
D. call on the learner regularly so he or she expects to be called upon and must be prepared for participation.

_____ **73.** A learner who uses a group situation to gain attention for him or herself is classified as:

A. a staller.
B. timid.
C. a troublemaker.
D. a show-off.

_____ **74.** The uninterested learner displays little ________________ or ________________ and has a low learning and accomplishment rate.

A. enthusiasm; I.Q.
B. common sense; good judgment
C. energy; attention
D. ability; desire

_____ **75.** If a skill is being taught, make sure the students have sufficient time:

A. to take notes.
B. to practice.
C. for breaks.
D. for verbal emphasis.

_____ **76.** Instructors should slow their speaking speed when:

A. beginning a class.
B. emphasizing important items.
C. students are taking notes.
D. All of the above

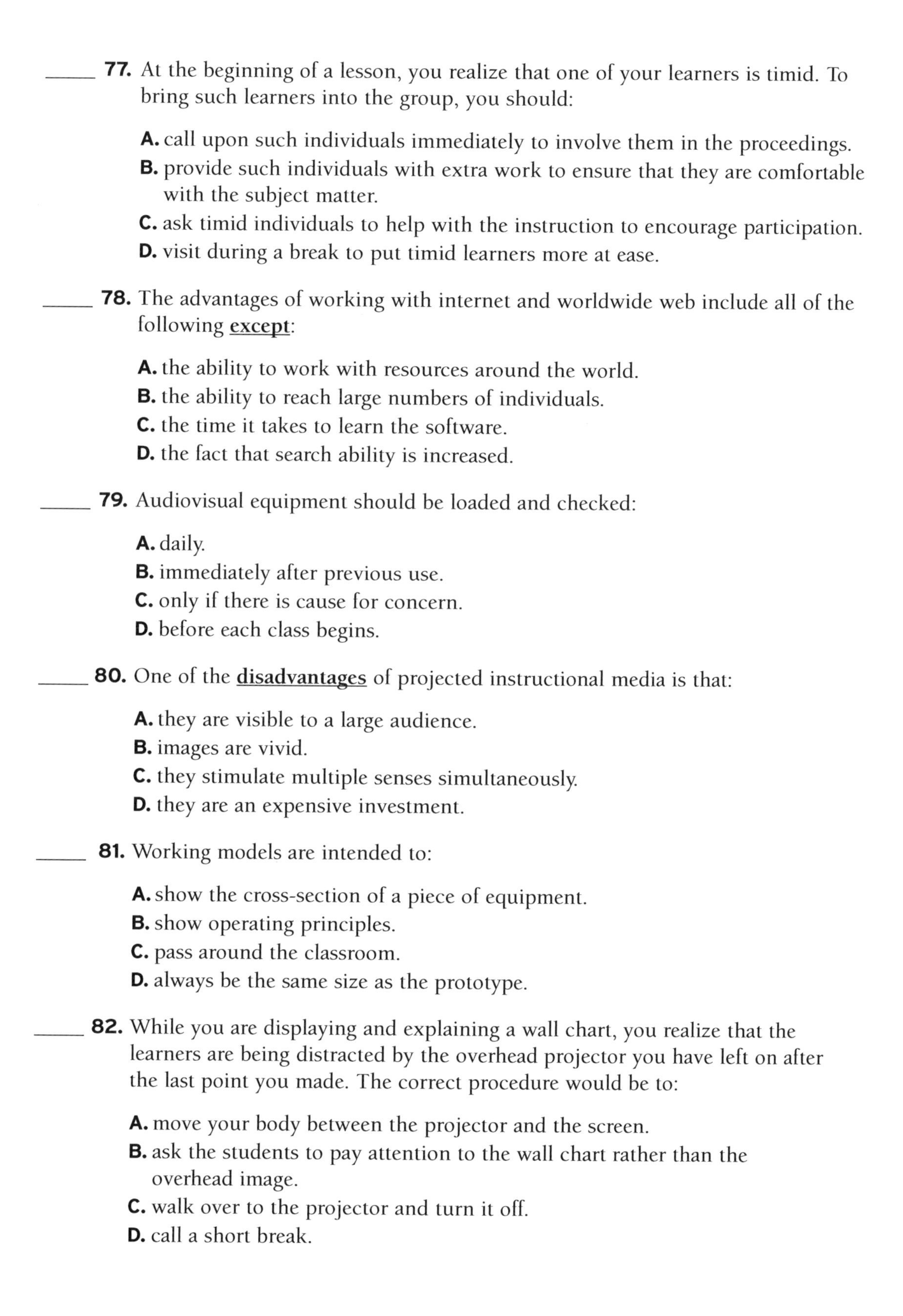

_____ **77.** At the beginning of a lesson, you realize that one of your learners is timid. To bring such learners into the group, you should:

A. call upon such individuals immediately to involve them in the proceedings.
B. provide such individuals with extra work to ensure that they are comfortable with the subject matter.
C. ask timid individuals to help with the instruction to encourage participation.
D. visit during a break to put timid learners more at ease.

_____ **78.** The advantages of working with internet and worldwide web include all of the following **except**:

A. the ability to work with resources around the world.
B. the ability to reach large numbers of individuals.
C. the time it takes to learn the software.
D. the fact that search ability is increased.

_____ **79.** Audiovisual equipment should be loaded and checked:

A. daily.
B. immediately after previous use.
C. only if there is cause for concern.
D. before each class begins.

_____ **80.** One of the **disadvantages** of projected instructional media is that:

A. they are visible to a large audience.
B. images are vivid.
C. they stimulate multiple senses simultaneously.
D. they are an expensive investment.

_____ **81.** Working models are intended to:

A. show the cross-section of a piece of equipment.
B. show operating principles.
C. pass around the classroom.
D. always be the same size as the prototype.

_____ **82.** While you are displaying and explaining a wall chart, you realize that the learners are being distracted by the overhead projector you have left on after the last point you made. The correct procedure would be to:

A. move your body between the projector and the screen.
B. ask the students to pay attention to the wall chart rather than the overhead image.
C. walk over to the projector and turn it off.
D. call a short break.

_____ **83.** Before using any training audiovisual aids, the **first** thing an instructor must do is:

A. introduce it properly to the group.
B. determine the best time to use the aid.
C. preview the audiovisual aid prior to class.
D. preview the audiovisual with a few of learners to see if it is relevant.

_____ **84.** A **disadvantage** of chalk/dry eraser marker boards is that:

A. they do not hold learners' attention.
B. once material is produced, it must be left in place or erased.
C. only highlights can be pointed out.
D. they require no special equipment.

_____ **85.** ________________ are convenient teaching aids because they provide a permanent record and are portable.

A. Chalk board
B. Dry erase boards
C. Electronic copy boards
D. Easel pads

_____ **86.** Training prerequisites are often used to determine participants for:

A. entry into training programs.
B. eligibility for hiring.
C. promotion.
D. All of the above

_____ **87.** To aid learners in reaching their highest potential in taking a test, the instructor should do all of the following **except**:

A. read instructions aloud while students read them silently.
B. provide directions that are clear.
C. give hints about individual items when answering questions.
D. keep instructions to a minimum during the test.

_____ **88.** A test that measures what it is supposed to measure is a ________________ test:

A. comprehensive
B. discriminating
C. valid
D. reliable

_____ **89.** ________________ tests may be used to supplement performance tests.

A. Oral
B. Written
C. Essay
D. Multiple choice

_____ **90.** The **primary** purpose of an evaluation process is to:

A. determine whether the instruction was adequate.
B. improve the teaching/learning process.
C. determine the level of learner insight.
D. allow the learners to express their viewpoints.

_____ **91.** Which of the following **is not** a purpose of written tests?

A. To find out whether students can perform the job taught
B. To provide the only means of evaluation
C. To identify student weaknesses
D. To hold learners accountable for material taught

_____ **92.** Most cheating on tests is the result of:

A. copying during test administration.
B. hints given by test administrators.
C. advance knowledge of test content.
D. Both A and C are correct

_____ **93.** Coaching includes all of the following **except**:

A. observation.
B. evaluation.
C. providing written directions.
D. suggestions for improvement.

_____ **94.** Under the Family Educational and Privacy Act of 1974 an individual's name and _________________ cannot be displayed **together**.

A. attendance
B. grades
C. age
D. marital status

_____ **95.** Training records that contain names, grading, and scoring information are subject to the provisions of:

A. the Family Leave Act of 1994.
B. Occupational Safety and Health Act.
C. the Family Education Rights and Privacy Act of 1974.
D. National Fire Protection Association Standards.

_____ **96.** When is a summative evaluation process employed?

A. Before the course
B. Periodically during course development
C. After the course
D. Continuously

_____ **97.** If a class scores poorly on a test, the instructor should:

A. throw out the poor test items (questions).
B. reteach the lesson.
C. change the distractors and give the test again.
D. revise the lesson plan.

_____ **98.** Which of the following is not considered a primary means of helping learners with learning disabilities?

A. Providing feedback on progress
B. Teaching to the slowest learning level
C. Individualized instruction
D. Tutoring

_____ **99.** A type of evaluation that can be used throughout a course delivery and provides immediate feedback is a _________________ evaluation.

A. comprehensive
B. formative
C. prescriptive
D. summative

_____ **100.** An instructor can gauge how thoroughly the learners are grasping the lesson by:

A. the success/fail rate of the learners.
B. the feedback received from learners.
C. evaluations by other instructors.
D. Both A and B are correct

Did you score higher than 80% on Examination I-3? Circle Yes or No in ink.

Feedback Step

Now, what do we do with your "yes" and "no" answers given throughout the SAEP process? First, return to any response that has "no" circled. Go back to the highlighted answers for those examination items missed. Read and study the paragraph preceding the location of the answer, as well as the paragraph following the paragraph where the answer is located. This will expand your knowledge base for the missed question, put it in a broader perspective, and improve associative learning. Remember, you are trying to develop mastery of the required knowledge. Scoring 80 percent on an examination is good, but it is not mastery performance. To be at the top of your group, you must score much higher than 80 percent on your training, promotion, or certification examination.

Carefully review the Summary of Key Rules for Taking an Examination and Summary of Helpful Hints on the next two pages. Do this review now and at least two additional times prior to taking your next examination.

Helpful Hint

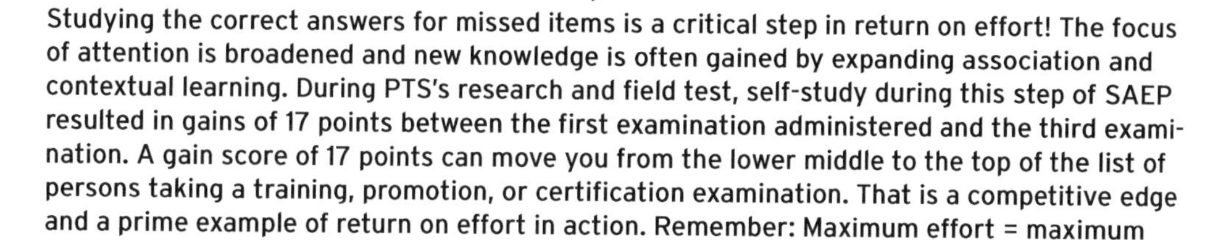

Studying the correct answers for missed items is a critical step in return on effort! The focus of attention is broadened and new knowledge is often gained by expanding association and contextual learning. During PTS's research and field test, self-study during this step of SAEP resulted in gains of 17 points between the first examination administered and the third examination. A gain score of 17 points can move you from the lower middle to the top of the list of persons taking a training, promotion, or certification examination. That is a competitive edge and a prime example of return on effort in action. Remember: Maximum effort = maximum results!

Summary of Key Rules for Taking an Examination

Rule 1—Examination preparation is not easy. Preparation is 95% perspiration and 5% inspiration.

Rule 2—Follow the steps very carefully. Do not try to reinvent or shortcut the system. It really works just as it was designed to!

Rule 3—Mark with an "X" any examination items for which you guessed the answer. For maximum return on effort, you should also research any answer that you guessed, even if you guessed correctly. Find the correct answer, highlight it, and then read the entire paragraph that contains the answer. Be honest and mark all questions on which you guessed. Some examinations have a correction for guessing built into the scoring process. The correction for guessing can reduce your final examination score. If you are guessing, you are not mastering the material.

Rule 4—Read questions twice if you have any misunderstanding, especially if the question contains complex directions or activities.

Rule 5—If you want someone to perform effectively and efficiently on the job, the training and testing program must be aligned to achieve this result.

Rule 6—When preparing examination items for job-specific requirements, the writer must be a subject matter expert with current experience at the level that the technical information is applied.

Rule 7—Good luck = good preparation.

Summary of Helpful Hints

Helpful Hint - Most of the time your first impression is the best. More than 41% of changed answers during PTS's SAEP field test were changed from a right answer to a wrong answer. Another 33% were changed from a wrong answer to another wrong answer. Only 26% of answers were changed from wrong to right. In fact three participants did not make a perfect score of 100% because they changed one right answer to a wrong one! Think twice before you change your answer. The odds are not in your favor.

Helpful Hint - Researching correct answers is one of the most important activities in SAEP. Locate the correct answer for all missed examination items. Highlight the correct answer. Then read the entire paragraph containing the answer. This will put the answer in context for you and provide important learning by association.

Helpful Hint - Proceed through all missed examination items using the same technique. Reading the entire paragraph improves retention of the information and helps you develop an association with the material and learn the correct answers. This step may sound simple. A major finding during the development and field testing of SAEP was that you learn from your mistakes.

Helpful Hint - Follow each step carefully to realize the best return on effort. Would you consider investing your money in a venture without some chance of earning a return on that investment? Examination preparation is no different. You are investing time and expecting a significant return for that time. If, indeed, time is money, then you are investing money and are due a return on that investment. Doing things right and doing the right things in examination preparation will ensure the maximum return on effort.

Helpful Hint - Try to determine why you selected the wrong answer. Usually something influenced your selection. Focus on the difference between your wrong answer and the correct answer. Carefully read and study the entire paragraph containing the correct answer. Highlight the answer.

Helpful Hint - Studying the correct answers for missed items is a critical step in achieving your desired return on effort! The focus of attention is broadened, and new knowledge is often gained by expanding association and contextual learning. During PTS's research and field test, self-study during this step of SAEP resulted in gains of 17 points between the first examination administered and the third examination. A gain score of 17 points can move you from the lower middle to the top of the list of persons taking a training, promotion, or certification examination. That is a competitive edge and a prime example of return on effort in action. Remember: Maximum effort = maximum results!

PHASE II

Fire Instructor II

Examination II-1, Surveying Weaknesses

At this point in SAEP, you should have the process of self-directed learning using examinations fixed in your mind. Moving through Phase II is accomplished in the same way as for Phase I. Do not attempt to skip steps in the process—after all, you now understand how SAEP works. Skipping steps can lead to a weak examination preparation result. The examinations will be more difficult in Phase II because of the increased level of required knowledge and skills. You will find that the SAEP methods move you gradually from the simple to the complex.

Do not study prior to taking the examination. Examination II-1 is designed to identify your weakest areas in terms of NFPA Standard 1041, Fire Instuctor II. Some steps in SAEP will require self-study of specific reference materials. Remove Examination II-1 from the book.

Mark all answers in ink to ensure that no corrections or changes are made later. Do not mark through answers or change answers in any way once you have selected the answer. Doing so indicates uncertainty regarding the answer. Mastery is not compatible with uncertainty.

Step 1—Take Examination II-1. When you have completed Examination II-1, go to Appendix B and compare your answers with the correct answers. Notice that each answer has reference materials with page numbers. If you missed the correct answer to the examination item, you have a source for conducting your correct answer research.

Step 2—Score Examination II-1. How many examination items did you miss? Write the number of missed examination items in the blank in ink. ________________ Enter the number of examination items you guessed in this blank. ________________ Enter these numbers in the designated locations on your Personal Progress Plotter.

Step 3—Now the learning begins! Carefully research the page cited in the reference material for the correct answer. For instance, if you are using IFSTA, *Fire and Emergency Services Instructor, Sixth Edition*, go to the page number provided and find the answer.

Rule 3

Mark with an "X" any examination items for which you guessed the answer. For maximum return on effort, you should also research any answer that you guessed, even if you guessed correctly. Find the correct answer, highlight it, and then read the entire paragraph that contains the answer. Be honest and mark all questions on which you guessed. Some examinations have a correction for guessing built into the scoring process. The correction for guessing can reduce your final examination score. If you are guessing, you are not mastering the material.

Rule 4

Read questions twice if you have any misunderstanding, especially if the question contains complex directions or activities.

Helpful Hint

Proceed through all missed examination items using the same technique. Reading the entire paragraph improves retention of the information and helps you develop an association with the material and correct answers. This step may sound simple. A major finding during the development and field testing of SAEP was that you learn from your mistakes.

Examination II-1

Directions

Remove Examination II-1 from the manual. First, take a careful look at the examination. There should be 75 examination items. Notice that a blank line precedes each examination item number. This line is provided for you to enter the answer to the examination item. Write the answer in ink. Remember the rule about not changing your answers. Our research has shown that changed answers are often incorrect, and, more often than not, the answer that is chosen first is correct.

If you guess the answer to a question, place an "X" or a check mark by your answer. This step is vitally important as you gain and master knowledge. We will explain how we treat the "guessed" items later in SAEP.

Take the examination. Once you complete it, go to Appendix B and score your examination. Once the examination is scored, carefully follow the directions for feedback on the missed and guessed examination items.

_____ **1.** When developing a training schedule, the scheduler should:

A. schedule as much training as possible, realizing most of it will get cancelled.
B. schedule one session at a time to ensure personnel attendance.
C. leave some open time so the schedule can be modified without affecting the entire year.
D. require the instructors to prepare on their own time.

_____ **2.** In the Five-Step Planning Process Model for Training Managers, the Identification Step includes which of the following activities:

A. conducting a strategic needs assessment.
B. conducting a strategic risk analysis.
C. identifying an instructor cadre.
D. reassessing strategic needs.
E. monitoring and modifying the program.

_____ **3.** Tracking training schedules, handling cancellations, and rescheduling training sessions should:

A. be given to one individual.
B. not occur if properly planned in advance.
C. be performed by a paid full-time instructor.
D. be rotated on a regular basis.

_____ **4.** It is an Instructor II's responsibility to ensure that the instructional team remains current. Professional development can be achieved by:

A. attending seminars.
B. attending workshops.
C. networking with other instructors.
D. All of the above

_____ **5.** Which of the following is **true** concerning a needs analysis?

A. It can be performed in a relatively short time frame.
B. It does not form an opinion about the target audience.
C. It is dependent upon existing materials and resources.
D. It includes a proposed budget for the program.

_____ **6.** Scheduling critical training is achievable by:

A. performing a needs analysis.
B. concentrating on minimal acceptable standards.
C. requiring fewer personnel to perform more duties.
D. conducting more public service programs.

_____ **7.** The ________________ budget category allows for expenses required for normal day-to-day functions of a training program.

A. operations
B. capital goods
C. zero-base
D. salaries and benefits

_____ **8.** The training needs of an organization should:

A. be limited to low cost items.
B. only reflect the desires of the Chief of the Department.
C. be driving the request in the budget process.
D. be directed by the Financial Manager.

_____ **9.** Once a budget has been determined, it is important to do all of the following **except**:

A. track expenditures.
B. report expenditures.
C. monitor revenues.
D. monitor all expenditures at year-end.

_____ **10.** Which item listed below is a capital expenditure?

A. Office supplies
B. Building
C. Heating and air conditioning
D. Insurance

_____ **11.** According to the five principles for budget management, the Instructor II should:

A. base the training budget on program needs.
B. summarize actual expenditures and revenues.
C. determine what programs may be offered, depending on the size of the budget.
D. use alternative funds before using fixed budgetary funds.

_____ **12.** _________________ is a broad, comprehensive term describing a person's or organization's responsibility under the law.

A. Guideline
B. Standard
C. Regulation
D. Liability

_____ **13.** A policy is a guiding principle that organizations use to:

A. identify a general philosophy.
B. address specific issues or problems.
C. require a step-by-step outline of a task.
D. provide specific rules and regulations.

_____ **14.** For documentation purposes, many departments require multiple copy forms that list procedures, rules, and other requirements to be signed by the:

A. supervisor.
B. learner.
C. training instructor.
D. chief.

_____ **15.** To maintain confidentiality of learner test records, the department should:

A. use a local area network system.
B. link the computer to the internet.
C. isolate all grades to a single floppy disk.
D. regulate system security by policy.

_____ **16.** If an instructor could not perform because there was no directive, what type of liability is this?

A. Foreseeability
B. Personal Liability
C. Vicarious Liability
D. Immunity

_____ **17.** When selecting effective instructors, a training manager will look for all of the following **except**:

A. credibility with the personnel being trained.
B. ability to "teach it all" rather than specialize.
C. technical proficiency in the subject matter.
D. roles held within the organization.

_____ **18.** When evaluating the performance of instructors, the evaluator should note whether the instructors:

A. worked within their skill level.
B. disclosed personal information.
C. presumed learner comprehension.
D. exceeded protocols to enhance learning.

_____ **19.** According to Maslow's Hierarchy of Needs, _______________ needs of instructors are satisfied when they are recognized by their peers or they are praised in front of their peers.

A. security
B. self-actualization
C. social
D. self-esteem

_____ **20.** An Instructor II should perform instructor evaluations for the following purposes **except** to:

A. ensure that instructors achieve training objectives.
B. improve overall instructor performance and quality.
C. find ways to decrease the training budget.
D. demonstrate that quality instruction is important to the organization.

_____ **21.** When writing objectives, the _______________ describes how well the performance is to be accomplished.

A. course description
B. behavior
C. task analysis
D. standard

_____ **22.** The terms goals and objectives are often used interchangeably; however, these terms **are not** synonymous. Objectives:

A. identify teaching needs.
B. determine what needs are to be taught.
C. describe the end result.
D. give a time frame for completion.

_____ **23.** Which of the following **is not** a part of a lesson plan?

A. Record of attendance
B. List of references
C. Assignment for next class
D. List of needed materials

_____ **24.** At what cognitive level of learning would you be able to interpret and solve fireground problems?

A. Level One
B. Level Two
C. Level Three
D. Level Four

_____ **25.** A behavioral objective consists of the following components:

A. condition, situation, and behavior.
B. behavior, task, and condition.
C. behavior, standard, and time.
D. condition, criteria, and behavior.

_____ **26.** Which of the following **is** **not** a component of a behavioral objective?

A. Task
B. Motivation
C. Criteria
D. Condition

_____ **27.** A behavioral objective **must** be stated:

A. at the beginning of each chapter.
B. clearly and distinctly by each learner.
C. in easy-to-understand terminology using a standard format.
D. in terms of measurable performance.

_____ **28.** In an occupational analysis, the items to be accomplished to complete the tasks are:

A. jobs.
B. operations.
C. key points.
D. units.

_____ **29.** On a screen, an instructor creates an image of a fire hazard situation that creates the illusion of fire and smoke. This statement **best** describes a/an:

A. simulation.
B. overhead projection.
C. multimedia presentation.
D. video tape.

_____ **30.** _________________ use a combination of audiovisual materials.

A. Easel pads
B. Illustrations
C. Cutaways
D. Multimedia

_____ **31.** When using non-projected instructional media, advantages to using _________________ are that they are inexpensive, require little storage space, and provide a permanent record.

A. simulation aids
B. easel pads
C. dry easel boards
D. audio cassettes

_____ **32.** The task is a:

A. component of knowledge and skill in an occupation.
B. step which leads to another step.
C. description of what the learner is expected to do or the product or result.
D. career or professional category.

_____ **33.** To give all learners an opportunity to learn, instructors **must**:

A. expose learners to consistent teaching methods.
B. practice one standard learning style.
C. allow learners to expand on the learning methods.
D. understand that differences in learning are based on the ability of the instructor.

_____ **34.** When developing a lesson plan which uses only one portion of a video tape, the developer should include instructions to:

A. use fast forward to cue the video at the proper location during the class.
B. mute the sound while fast forwarding during the class.
C. cue the tape to the proper location prior to the start of class.
D. Both A and B are correct

_____ **35.** Which of the following example statements **would not be** suitable for use when writing behavioral objectives?

A. Identify the pike pole in the illustration.
B. Appreciate the history of the fire service.
C. Describe the procedure for folding salvage covers.
D. Demonstrate taking the roof ladder from the apparatus and putting it on the roof.

_____ **36.** In the development of a lesson plan, mock-ups are best suited for _______________ lessons.

A. lecture
B. illustration
C. demonstration
D. discussion

_____ **37.** To be effective, the fire instructor must utilize the three domains of learning. Which of the following is **true** concerning the three domains of learning?

A. For effective learning, the areas must overlap or interrelate with each other.
B. The domains are not related; each may be experienced separately without overlaps.
C. The cognitive and psychomotor domains are the what, how, and why of learning.
D. All of the above

_____ **38.** Learning objectives:

A. indicate what instructors should teach.
B. are written for instructors.
C. provide instructors with verbal requirements.
D. describe what the learner will accomplish.

_____ **39.** There are ________________ steps of instruction.

A. one
B. two
C. three
D. four

_____ **40.** The purpose of a/an ________________ is to provide the instructor the necessary detail with which to teach a job, including both psychomotor skills and required knowledge.

A. job breakdown sheet
B. behavioral objective
C. lesson summary
D. evaluation

_____ **41.** During the development of the lesson plan, the instructional step that should be used to establish lesson relevancy is:

A. preparation.
B. presentation.
C. application.
D. evaluation.

_____ **42.** A form used to provide the learner with opportunities to use steps or use multiple skills to complete an activity is the:

A. study sheet.
B. job breakdown sheet.
C. information sheet.
D. task worksheet.

_____ **43.** In order to effectively handle low literacy levels, an instructor should:

A. use a variety of materials instead of long lectures.
B. use technical terms and complicated vocabulary to enhance learner literacy.
C. expect the same application from low literacy learners as the majority of the class.
D. tailor class to the strongest student and not weakest, so low literacy learners have to strive.

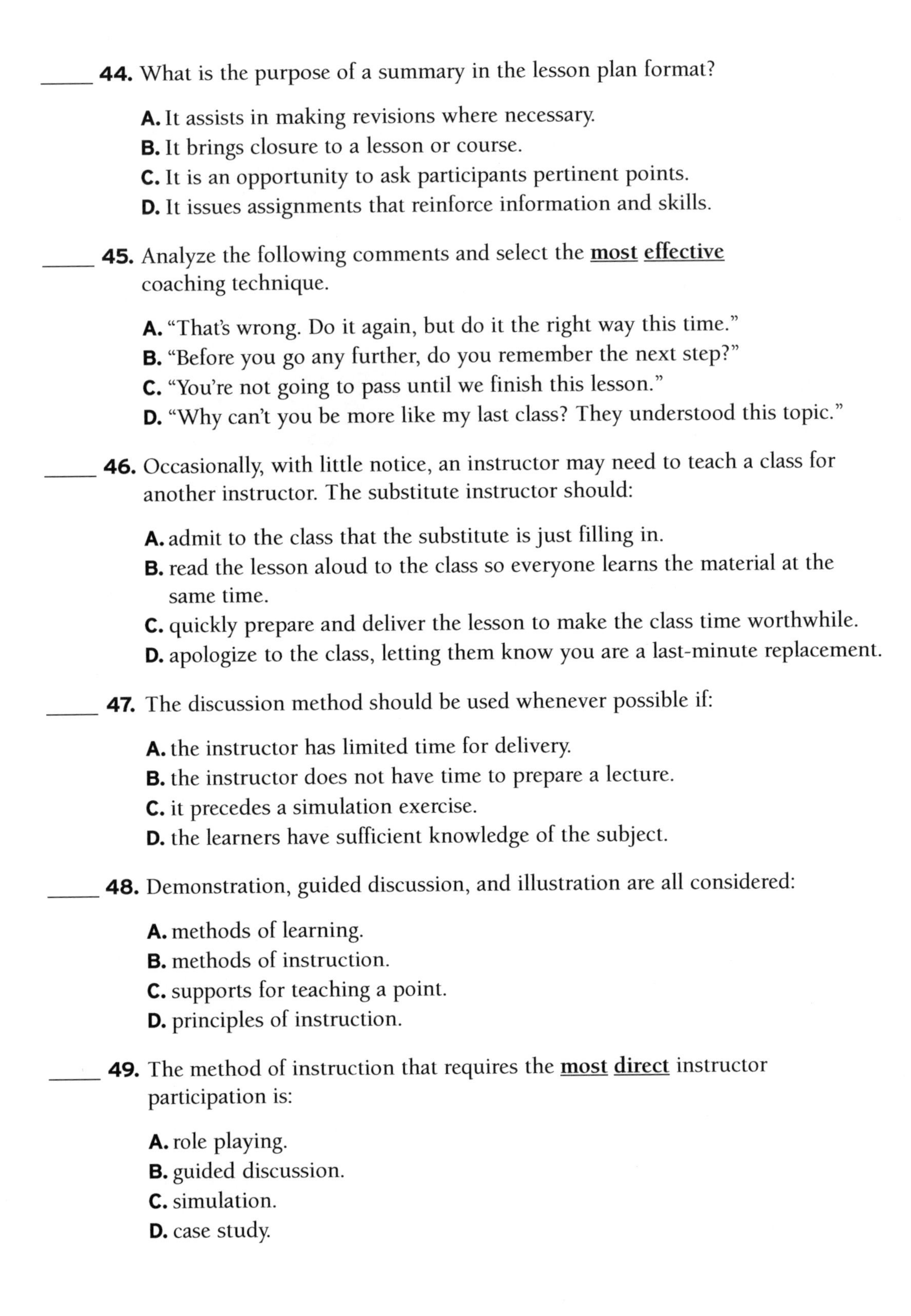

____ **44.** What is the purpose of a summary in the lesson plan format?

A. It assists in making revisions where necessary.
B. It brings closure to a lesson or course.
C. It is an opportunity to ask participants pertinent points.
D. It issues assignments that reinforce information and skills.

____ **45.** Analyze the following comments and select the **most effective** coaching technique.

A. "That's wrong. Do it again, but do it the right way this time."
B. "Before you go any further, do you remember the next step?"
C. "You're not going to pass until we finish this lesson."
D. "Why can't you be more like my last class? They understood this topic."

____ **46.** Occasionally, with little notice, an instructor may need to teach a class for another instructor. The substitute instructor should:

A. admit to the class that the substitute is just filling in.
B. read the lesson aloud to the class so everyone learns the material at the same time.
C. quickly prepare and deliver the lesson to make the class time worthwhile.
D. apologize to the class, letting them know you are a last-minute replacement.

____ **47.** The discussion method should be used whenever possible if:

A. the instructor has limited time for delivery.
B. the instructor does not have time to prepare a lecture.
C. it precedes a simulation exercise.
D. the learners have sufficient knowledge of the subject.

____ **48.** Demonstration, guided discussion, and illustration are all considered:

A. methods of learning.
B. methods of instruction.
C. supports for teaching a point.
D. principles of instruction.

____ **49.** The method of instruction that requires the **most direct** instructor participation is:

A. role playing.
B. guided discussion.
C. simulation.
D. case study.

_____ **50.** Effective handling of a question includes all of the following **except**:

A. saying, "I don't know, but I'll find out."
B. answering the question thoroughly and in depth when the question is of interest to only the person asking.
C. referring the learner to related information when there is no exact answer to the question.
D. when a question refers to material that is covered in a later lesson, answering the question briefly and explaining that it will be covered in more detail later.

_____ **51.** During the _________________ step of the instructional delivery, the instructor should get the learners' attention, arouse curiosity, and develop interest.

A. lecture
B. preparation
C. application
D. demonstration

_____ **52.** In a situation with live-burn training and fifteen learners, the Incident Command System instructor must be sure that all other instructors:

A. have as many learners working as possible.
B. are qualified to teach and supervise learners.
C. rotate with the all the learners from station to station.
D. know they can improvise and change the situation as may be needed.

_____ **53.** Protective policies such as worker's compensation acts are examples of _________________ law.

A. common
B. administrative
C. judiciary
D. statutory

_____ **54.** By regularly conducting an occupational or job analysis of a fire fighter, the fire instructor is:

A. reducing vicarious liability.
B. practicing foreseeability.
C. justifying fireground injuries.
D. performing limited immunity.

_____ **55.** Which of the following **does not** constitute high-hazard training?

A. Use of power tools
B. Confined space rescue
C. Hazardous materials
D. CPR mannequin practice

____ **56.** The instructor should ensure that water supply operations during high-hazard training, such as live burns:

A. are reliable for the entire duration of the exercise.
B. give students practical on-the-job training.
C. are remote from the incident to allow practice in relay operations.
D. are performed by certified fire fighters only.

____ **57.** Tanks used for training in confined-space rescue should:

A. expose the student to the same dangers expected in a real emergency.
B. have only one method for both access and egress.
C. contain simulated smoke to limit visibility.
D. have one side or end cut away to facilitate a safer training environment.

____ **58.** It is ________________ to burn buildings in order to practice search and rescue techniques.

A. cost effective
B. realistic
C. expensive
D. unsafe

____ **59.** A thunderstorm occurs midway through an outdoor practical drill. You should:

A. continue with the drill. Safety is secondary to skill development.
B. discontinue the drill. Learners will miss this portion of the program.
C. discontinue the drill until the inclement weather has passed.
D. continue the drill since bad weather conditions are a part of firefighting.

____ **60.** ________________ tests are used to determine the level of knowledge and readiness required when determining placement in a course.

A. Prescriptive
B. Classification
C. Performance
D. Progress evaluation

____ **61.** A/an ________________ test requires learners to analyze, revise, redesign, or evaluate a problem.

A. objective
B. subjective
C. oral
D. performance

_____ **62.** To identify learner mastery or non-mastery of subject matter, a _______________ test should be administered.

A. criterion-referenced
B. diagnosis
C. norm-referenced
D. classification

_____ **63.** One <u>**disadvantage**</u> of a multiple choice test is that it:

A. cannot measure a variety of types of knowledge.
B. cannot measure complex learning outcomes.
C. is difficult to write plausible distractors.
D. is difficult to diagnose results.

_____ **64.** A test designed to measure learner improvement throughout a course or unit of instruction is a _______________ test.

A. progress
B. summative
C. standardized
D. norm-referenced

_____ **65.** The correct choice in a multiple choice test is referred to as the answer; the remaining choices are called:

A. negatives.
B. positives.
C. stems.
D. distractors.

_____ **66.** To be _______________, a test should accurately and consistently evaluate performance.

A. valid
B. comprehensive
C. difficult
D. reliable

_____ **67.** It is important to use the same type of equipment when testing and practicing because:

A. it maintains continuity in learning sessions.
B. it makes testing valid and reliable.
C. learners should train on the same or generically similar equipment as on the job.
D. All of the above

____ **68.** In evaluating a course, it is critical to make clear to anyone involved ________________ is to be evaluated and ________________.

A. what; why
B. who; when
C. what; where
D. who; how

____ **69.** The primary purpose of course evaluation is to provide the instructor with feedback to improve the ________________ process.

A. teaching/learning
B. discussion
C. skill/demonstration
D. testing

____ **70.** To evaluate a pilot program, developers, and instructors perform ________________ evaluations after the program ends.

A. summative
B. formative
C. process
D. survey

____ **71.** Evaluation instruments provide:

A. feedback on test validity and reliability.
B. the opportunity for learners to practice their skills.
C. clear job performance objectives.
D. identification of strategic goals.

____ **72.** A test item that measures what it is supposed to measure is:

A. discriminating.
B. valid.
C. reliable.
D. comprehensive.

____ **73.** To be ________________, a test must be constructed so it can measure a learner's achievement over all phases of the course.

A. realistic
B. comprehensive
C. progressive
D. valid

_____ **74.** A test that separates learners who know the material from those who do not is said to be:

A. ambiguous.
B. comprehensive.
C. valid.
D. discriminating.

_____ **75.** In test analysis, raw score consists of:

A. points learner receives on a test.
B. percentage score times median score.
C. median score plus the mean.
D. percentage score added to the mean.

Did you score higher than 80% on Examination II-1? Circle Yes or No in ink. (We will return to your answer to this question later in SAEP.)

Examination II-2, Adding Difficulty and Depth

During Examination II-2, progress will be made in developing your depth of knowledge and skills. Reminder: Follow the steps carefully to realize the best return on effort.

Step 1—Take Examination II-2. When you have completed Examination II-2, go to Appendix B and compare your answers with the correct answers.

Step 2—Score Examination II-2. How many examination items did you miss? Write the number of missed examination items in the blank in ink. ________________ Enter the number of examination items you guessed in this blank. ________________ Enter these numbers in the designated locations on your Personal Progress Plotter.

Step 3—Once again the learning begins. During the feedback step, use the Appendix B information for Examination II-2 to research the correct answers for items you missed or guessed. Highlight the correct answer during your research of the reference materials. Read the entire paragraph containing the correct answer.

Examination II-2

Directions

Remove Examination II-2 from the manual. First, take a careful look at the examination. There should be 75 examination items. Notice that a blank line precedes each examination item number. This line is provided for you to enter the answer to the examination item. Write the answer in ink. Remember the rule about not changing your answers. Our research has shown that changed answers are often incorrect, and, more often than not, the answer that is chosen first is correct.

If you guess the answer to a question, place an "X" or a check mark by your answer. This step is vitally important as you gain and master knowledge. We will explain how we treat the "guessed" items later in SAEP.

Take the examination. Once you complete it, go to Appendix B and score your examination. Once the examination is scored, carefully follow the directions for feedback on the missed and guessed examination items.

_____ **1.** The Instructor II duties include:

A. establishing general departmental policies and procedures.
B. evaluating department company officers during fire ground operations.
C. program scheduling.
D. All of the above

_____ **2.** An Instructor II's responsibility is to ensure that the instructional team members realize that their primary role is to:

A. set organizational policies.
B. elicit feedback from learners and colleagues.
C. plan, develop, and conduct training sessions.
D. ensure that learner participation is encouraged.

_____ **3.** In the Five-Step Planning Process Model for Training Managers, the Selection Step includes which of the following activities:

A. conduct a strategic needs assessment.
B. conduct a strategic risk analysis.
C. identify an instructor cadre.
D. reassess strategic needs.

_____ **4.** In the Five-Step Planning Process Model for Training Managers, the Implementation Step includes which of the following activities:

A. conducting a strategic needs assessment.
B. conducting a strategic risk analysis.
C. identifying an instructor cadre.
D. reassessing strategic needs.
E. monitoring and modifying the program.

_____ **5.** When formulating budget needs, program developers must estimate the cost and benefits of the course. This is also known as the _______________ component in the planning process model.

A. identification
B. selection
C. design
D. evaluation

_____ **6.** A/An _______________ budget generally categorizes requests into salaries and benefits, operations, and capital goods.

A. zero-base
B. training
C. integrative
D. performance

_____ **7.** Line-item accounting is a form of:

A. bargaining.
B. complaint resolution.
C. budgetary control.
D. an employee accountability system.

_____ **8.** Which of the following is/are method(s) used to justify preliminary budget figures?

A. Identifying fixed and recurring expenses
B. Letting officials know how spending now will bring future savings
C. Presenting little information and forcing officials to ask questions
D. Both A and B are correct.

_____ **9.** Which item listed below is a capital expenditure?

A. Office supplies
B. Building
C. Heating and air conditioning
D. Insurance

_____ **10.** Budget expenses categorized as human resources include:

A. worker's compensation.
B. office furniture.
C. office machines.
D. communication services.

_____ **11.** Managing funding and resources to achieve training goals can be accomplished through which of the following nontraditional methods?

A. Developing cooperative relationships with the industry
B. Seeking funding through grants
C. Raising funds by providing training to the private sector
D. All of the above

_____ **12.** The term that embodies the concept that an individual's records are confidential is:

A. contractual agreement.
B. right of privacy.
C. employee rights.
D. agreement act.

_____ **13.** What legal act prevents disclosure of personal information without consent?

A. The American with Disabilities Act of 1980
B. The Civil Rights Act of 1964
C. Equal Employment Opportunity Act of 1974
D. Family Educational Rights and Privacy Act of 1974

_____ **14.** A policy is a guiding principle that organizations use to:

A. identify a general philosophy.
B. address specific issues or problems.
C. require a step-by-step outline of a task.
D. provide specific rules and regulations.

_____ **15.** Programs designed to seek persons of minority races and ethnic backgrounds are known as:

A. Operation Outreach
B. Affirmative Action
C. Equal Employment Opportunities
D. Right-to-Work

_____ **16.** Employment programs that are required by federal statutes designed to correct discriminatory practices in hiring minority group members are the:

A. equal employment opportunity laws.
B. Americans with Disabilities Acts.
C. affirmative action programs.
D. Hiring Fairness Acts of 1989.

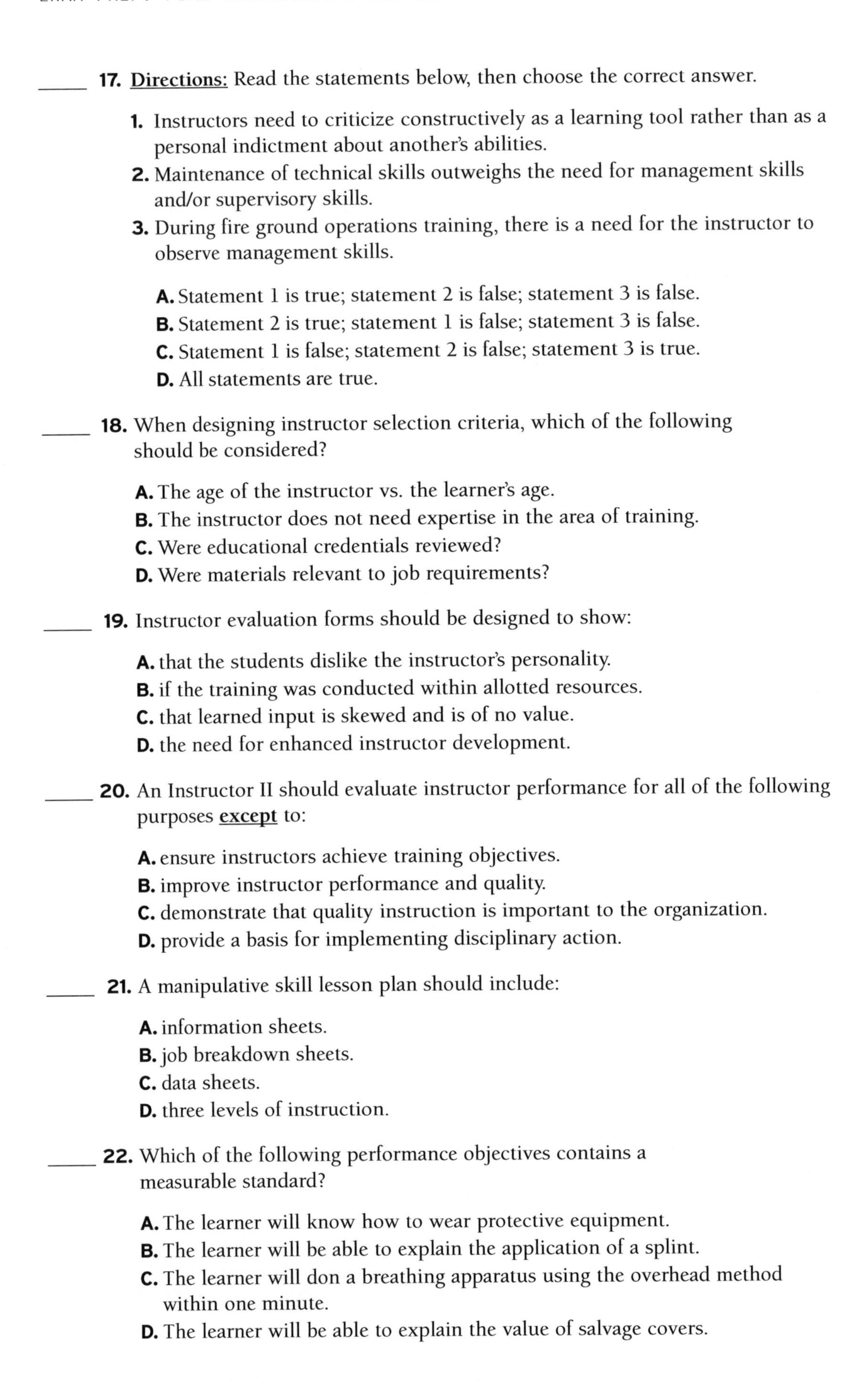

_____ **17.** **Directions:** Read the statements below, then choose the correct answer.

1. Instructors need to criticize constructively as a learning tool rather than as a personal indictment about another's abilities.
2. Maintenance of technical skills outweighs the need for management skills and/or supervisory skills.
3. During fire ground operations training, there is a need for the instructor to observe management skills.

A. Statement 1 is true; statement 2 is false; statement 3 is false.
B. Statement 2 is true; statement 1 is false; statement 3 is false.
C. Statement 1 is false; statement 2 is false; statement 3 is true.
D. All statements are true.

_____ **18.** When designing instructor selection criteria, which of the following should be considered?

A. The age of the instructor vs. the learner's age.
B. The instructor does not need expertise in the area of training.
C. Were educational credentials reviewed?
D. Were materials relevant to job requirements?

_____ **19.** Instructor evaluation forms should be designed to show:

A. that the students dislike the instructor's personality.
B. if the training was conducted within allotted resources.
C. that learned input is skewed and is of no value.
D. the need for enhanced instructor development.

_____ **20.** An Instructor II should evaluate instructor performance for all of the following purposes **except** to:

A. ensure instructors achieve training objectives.
B. improve instructor performance and quality.
C. demonstrate that quality instruction is important to the organization.
D. provide a basis for implementing disciplinary action.

_____ **21.** A manipulative skill lesson plan should include:

A. information sheets.
B. job breakdown sheets.
C. data sheets.
D. three levels of instruction.

_____ **22.** Which of the following performance objectives contains a measurable standard?

A. The learner will know how to wear protective equipment.
B. The learner will be able to explain the application of a splint.
C. The learner will don a breathing apparatus using the overhead method within one minute.
D. The learner will be able to explain the value of salvage covers.

_____ **23.** Given a length of 2 ½ inch hose and a fog nozzle, the learner will connect the nozzle to the hose using the over-the-hip method completing all steps with 100% accuracy within 10 seconds. The underlined portion of the preceding statement represents the:

A. condition.
B. behavior.
C. standard.
D. application.

_____ **24.** An Instructor II is designing a new training program. The first step is to:

A. recruit and select instructors.
B. prioritize training topics.
C. determine resources required.
D. identify the purpose of the training program.

_____ **25.** The part of a behavioral objective that describes the tools or equipment a learner is to use to complete a task is known as the:

A. standard.
B. condition.
C. application.
D. behavior.

_____ **26.** In developing a job breakdown sheet, begin by listing _______________; then list the _______________ for performance that instructors **must stress** while teaching.

A. blocks; operations
B. operations; key points
C. operations; blocks
D. tasks; operations

_____ **27.** Which of the following is not part of an occupational analysis?

A. Unit
B. Block
C. Evaluation
D. Task

_____ **28.** An **important** consideration when determining the need for a training aid is:

A. to support instruction and enhance learning.
B. whether it will be acceptable to the class.
C. the added time factor.
D. availability and cost factors.

_____ **29.** What is a **disadvantage** of an easel pad?

A. Distracting when on display continuously
B. Provides a limited writing space
C. May be time-consuming to prepare simulations on
D. It is not a supplement to a chalkboard or a dry erase board

_____ **30.** The teaching aid that enables learners to bring elements of reality to the instructional environment is known as:

A. illustration.
B. case study.
C. brainstorming.
D. simulation.

_____ **31.** When duplicating materials, how can one be certain a copyright **has not been** infringed upon?

A. Charge learners for copied materials
B. Attach a copy of the Fair Use Doctrine
C. Obtain written permission from the publisher/owner
D. Have a lawyer review all duplications

_____ **32.** In the psychomotor domain of learning, the instructor **must**:

A. punish poor or unacceptable performance.
B. assure that each learner is allowed the same time period to comprehend the subject.
C. quickly move from one psychomotor level to another.
D. understand learner abilities at each level.

_____ **33.** Outlines that map out the information and skills to be taught and state the format or method to be used in delivering the instruction are called:

A. objectives.
B. task analysis.
C. program development.
D. lesson plans.

_____ **34.** A needs analysis accomplishes all of the following **except**:

A. resolving on-the-job problems
B. analyzing training in relation to job requirements
C. identifying lack of equipment to perform job skills
D. determining whether training is required

_____ **35.** Which of the following is true concerning objectives?

A. They are the basis for testing.
B. They must be designed to measure abstract affective goals.
C. They have limited effect as a measurement tool.
D. They require an opinion rather than an observation.

_____ **36.** Lesson plans:

A. create complexity through standardization.
B. do not include information on resources required for implementation.
C. do not allow for individual instructor input.
D. verify that information presented is appropriate for testing.

_____ **37.** When modifying an existing lesson plan, instructors **must** start with the:

A. objectives.
B. preparation.
C. application.
D. evaluation.

_____ **38.** The task analysis component called a performance objective is a:

A. combination of jobs and duties.
B. step which leads to another step in a job.
C. description of what the learner is expected to do or the product or result of the doing.
D. career or professional category that contains verb statements.

_____ **39.** A job is a:

A. combination of duties.
B. step in a series which leads to another step in a job.
C. description of what the learner is expected to do or the product or result.
D. career or professional category.

_____ **40.** The task analysis component called an occupation is defined as a:

A. combination of jobs and duties.
B. step which leads to another step in a job.
C. description of what the learner is expected to do or the product or result of the doing.
D. career or professional category.

_____ **41.** In preparing to write behavioral objectives, the correct order of steps is:

A. identifying the specific task to be taught, writing the objective, choosing the level of instruction.
B. writing the objective, identifying the specific task to be taught, choosing the level of instruction.
C. identifying the specific task to be taught, choosing the level of instruction, writing the objective.
D. choosing the level of instruction, writing the objective, identifying the specific task to be taught.

_____ **42.** Uniformity in teaching is accomplished by:

A. having learners use reference books.
B. always having the same instructor teach a particular subject.
C. using the same audio/visual materials each time a subject is taught.
D. the use of a lesson plan.

_____ **43.** Certain assumptions can be made when teaching adults. These include:

A. adults need to be self-directed.
B. adults will learn what they need to know to meet job requirements.
C. adults' orientation is problem-centered.
D. All the above are correct

_____ **44.** In psychomotor learning, the coaching process includes all of the following **except**:

A. observation.
B. evaluation.
C. suggestions.
D. discipline.

_____ **45.** As a lead instructor, you have been asked to address a large number of people about a new concern. Because of time and budget constraints, which of the following methods would you use to make your presentation?

A. Lecture
B. Illustration
C. Demonstration
D. Discussion

_____ **46.** The discussion method should be used whenever possible if:

A. the instructor has limited time for delivery.
B. the instructor does not have time to prepare a lecture.
C. it precedes a simulation exercise.
D. the learners have sufficient knowledge of the subject.

_____ **47.** To conduct a successful conference, an instructor should:

A. allow spontaneous participation.
B. assist learners in realizing that their knowledge is limited.
C. provide clear direction toward a clearly stated end result.
D. emphasize the conference leader's knowledge of the subject.

_____ **48.** The ________________ reflects the final expected outcome of a lesson.

A. level of instruction
B. task to be accomplished
C. task content
D. self-satisfaction step

_____ **49.** An instructor can recapture the daydreamer's attention by:

A. asking direct questions.
B. diverting attention away from the daydreamer.
C. determining the cause of the inattention.
D. asking the daydreamer to pay attention.

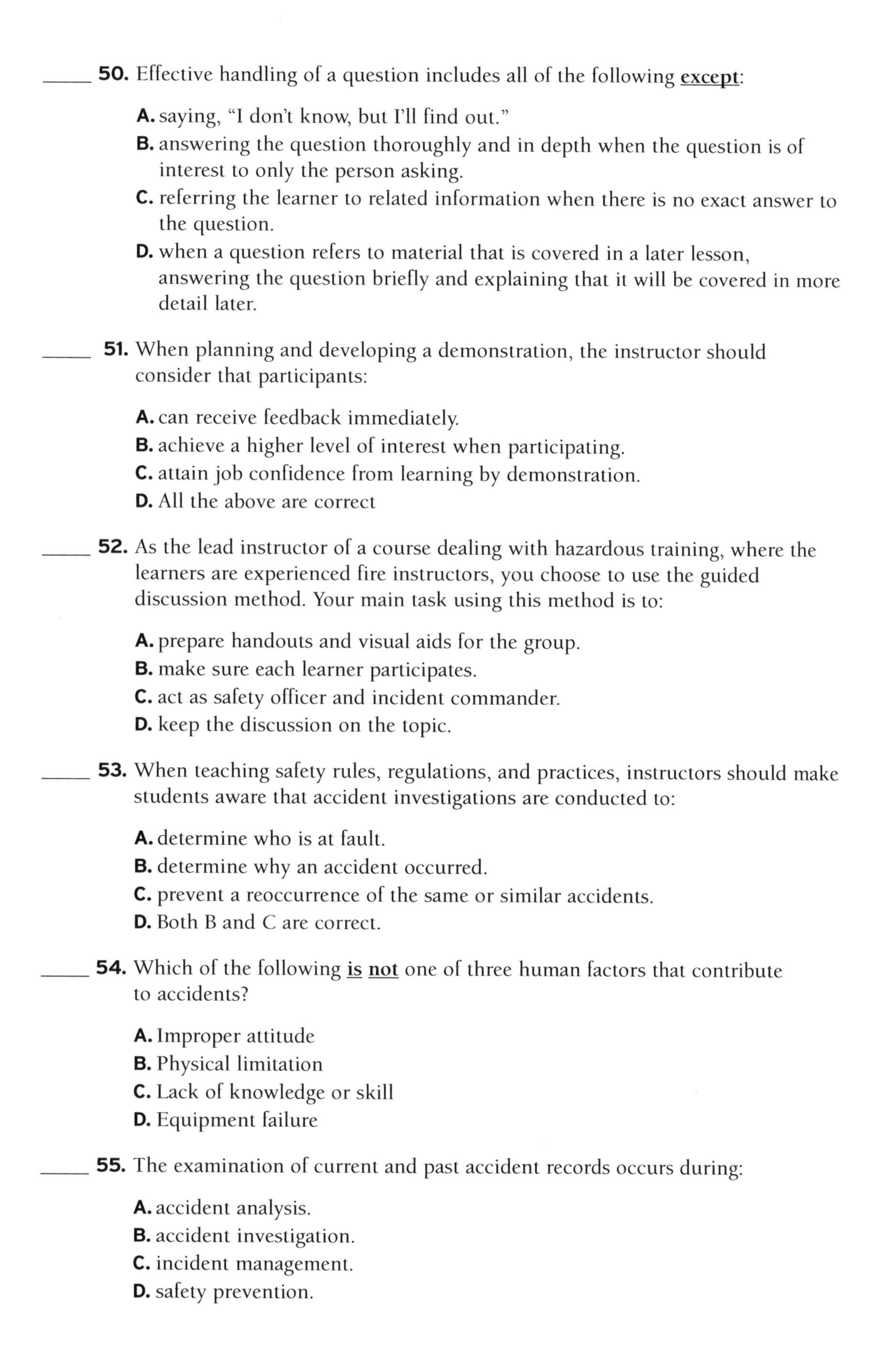

_____ **50.** Effective handling of a question includes all of the following **except**:

A. saying, "I don't know, but I'll find out."
B. answering the question thoroughly and in depth when the question is of interest to only the person asking.
C. referring the learner to related information when there is no exact answer to the question.
D. when a question refers to material that is covered in a later lesson, answering the question briefly and explaining that it will be covered in more detail later.

_____ **51.** When planning and developing a demonstration, the instructor should consider that participants:

A. can receive feedback immediately.
B. achieve a higher level of interest when participating.
C. attain job confidence from learning by demonstration.
D. All the above are correct

_____ **52.** As the lead instructor of a course dealing with hazardous training, where the learners are experienced fire instructors, you choose to use the guided discussion method. Your main task using this method is to:

A. prepare handouts and visual aids for the group.
B. make sure each learner participates.
C. act as safety officer and incident commander.
D. keep the discussion on the topic.

_____ **53.** When teaching safety rules, regulations, and practices, instructors should make students aware that accident investigations are conducted to:

A. determine who is at fault.
B. determine why an accident occurred.
C. prevent a reoccurrence of the same or similar accidents.
D. Both B and C are correct.

_____ **54.** Which of the following **is not** one of three human factors that contribute to accidents?

A. Improper attitude
B. Physical limitation
C. Lack of knowledge or skill
D. Equipment failure

_____ **55.** The examination of current and past accident records occurs during:

A. accident analysis.
B. accident investigation.
C. incident management.
D. safety prevention.

_____ **56.** The agency that develops standards to cover areas such as hazardous materials, protective equipment, and footwear is the:

A. Occupational Health and Safety Administration.
B. National Fire Administration.
C. American Society for Testing and Manufacturing.
D. National Institute for Occupational Safety and Health.

_____ **57.** Fire instructors must adhere to organizational safety rules and regulations that may include all of the following **except**:

A. risk management programs to control risk.
B. physical fitness requirements for learners only.
C. safety policies pertinent to local or national standards.
D. insurance policies for personnel in event of injury.

_____ **58.** Safety during high-hazard training means the instructor must ensure that:

A. students adhere to safety practices that instructor does not adhere to.
B. if learners are required to wear protective gear, then the instructor must also wear the same level of gear.
C. safety procedures are followed when convenient.
D. compliance does not effect attitude.

_____ **59.** A thunderstorm occurs midway through an outdoor practical drill. You should:

A. continue with the drill. Safety is secondary to skill development.
B. discontinue the drill. Learners will miss this portion of the program.
C. discontinue the drill until the inclement weather has passed.
D. continue the drill since bad weather conditions are a part of firefighting.

_____ **60.** A test used to measure an individual's proficiency in accomplishing a job or evolution is:

A. prescriptive.
B. progress.
C. multiple choice.
D. performance.

_____ **61.** A well-constructed ________________ test is generally recognized as one of the **most** versatile of the objective tests.

A. matching
B. multiple choice
C. true/false
D. essay

_____ **62.** ________________ tests are typically given in the middle or at the end of instruction.

A. Prescriptive
B. Application
C. Comprehensive
D. Both A and C are correct

_____ **63.** True/false, multiple choice, and matching tests are examples of _________________ tests.

A. objective
B. subjective
C. oral
D. performance

_____ **64.** A written test that **minimizes** the possibility of learner guessing is the:

A. short answer.
B. true/false.
C. multiple choice.
D. matching.

_____ **65.** An **advantage** of matching tests is that they:

A. minimize guessing.
B. measure complete understanding.
C. are easy to construct.
D. Both A and C are correct

_____ **66.** A written test may be:

A. difficult to administer.
B. manipulative.
C. subjective or objective.
D. performance-oriented.

_____ **67.** The **two most important** conditions of a well-designed test are:

A. validity and norm-referencing.
B. validity and reliability.
C. reliability and criterion-referencing.
D. consistency and accuracy.

_____ **68.** To be _________________, a test should accurately and consistently evaluate performance.

A. valid
B. comprehensive
C. difficult
D. reliable

_____ **69.** A final course evaluation feedback determines:

A. whether the instructional process has met the course objectives.
B. average learner scores.
C. learner participation in the course.
D. whether the course is needed or not.

_____ **70.** A _______________ evaluation is a post-course appraisal.

A. formative
B. practical
C. summative
D. All of the above

_____ **71.** _______________ evaluation is the ongoing, repeated checking during course development and instruction to determine the most effective instructional content, methods, and testing techniques.

A. Observation
B. Field testing
C. Summative
D. Formative

_____ **72.** In an evaluation effort, the three elements critical to measuring success are:

A. criteria, evidence, and judgment.
B. goals, objectives, and performance.
C. criteria, performance, and ability.
D. ability, judgment, and progress.

_____ **73.** <u>**Directions:**</u> Read the statements below, then choose the correct answer.

1. Methods of analyzing test results are generally referred to as "test statistics."
2. Statistics are a way of organizing, analyzing, and interpreting test scores.
3. Analyzing tests has no affect in determining test validity.

A. Statement 1 is true; statement 2 is false; statement 3 is false.
B. Statement 1 is true; statement 2 is true; statement 3 is false.
C. Statement 1 is false; statement 2 is true; statement 3 is false.
D. Statement 1 is false; statement 2 is false; statement 3 is true.

_____ **74.** Instructors who analyze student feedback may determine:

A. that students are biased and unable to recognize the instructor's depth of knowledge.
B. areas to adjust and improve.
C. low scores mean that the teacher did not teach.
D. students are unable to be objective.

_____ **75.** The purpose of test result analysis is to:

A. determine the validity and reliability of the test.
B. review behavior objectives.
C. offers students several scores.
D. show the difficulty of a test.

Did you score higher than 80% on Examination II-2? Circle Yes or No in ink. (We will return to your answer to this question later in SAEP.)

Examination II-3, Surveying Weaknesses and Improving Examination-Taking Skills

Examination II-3 section is designed to identify your remaining weaknesses in areas of NFPA 1041 for Fire Instructor II. This examination is randomly generated and contains examination items you have taken before, as well as new ones. There will be steps in the SAEP that require self-study of specific reference materials.

Mark all answers in ink to ensure that no corrections or changes are made later. Do not mark through answers or change answers in any way once you have selected your answer.

Step 1—Take Examination II-3. When you have completed Examination II-3, go to Appendix B and compare your answers with the correct answers. Notice that each answer has reference materials with page numbers. If you missed the correct answer to the examination item, you have a source for conducting your correct answer research.

Step 2—Score Examination II-3. How many examination items did you miss? Write the number of missed examination items in the blank in ink. ________________ Enter the number of examination items you guessed in this blank. ________________ Enter these numbers in the designated locations on your Personal Progress Plotter.

Step 3—Now you will begin reinforcing what you have learned! During the feedback step, research the correct answer using Appendix B information for Examination II-3. Highlight the correct answer during your research of the reference materials. Read the entire paragraph containing the correct answer.

Examination II-3

Directions

Remove Examination II-3 from the manual. First, take a careful look at the examination. There should be 100 examination items. Notice that a blank line precedes each examination item number. This line is provided for you to enter the answer to the examination item. Write the answer in ink. Remember the rule about not changing your answers. Our research has shown that changed answers are often incorrect, and, more often than not, the answer that is chosen first is correct.

If you guess the answer to a question, place an "X" or a check mark by your answer. This step is vitally important as you gain and master knowledge. We will explain how we treat the "guessed" items later in SAEP.

Take the examination. Once you complete it, go to Appendix B and score your examination. Once it is scored, carefully follow the directions for feedback on the missed and guessed examination items.

_____ **1.** Herzberg's theory of job enhancement states that:

A. problems with no resolution result in dissatisfaction.
B. satisfaction and motivation are not related to the level of participation.
C. two factors, motivation and stimulus, are responsible for job satisfaction.
D. job satisfaction and performance are not related to participation.

_____ **2.** The Instructor II duties include:

A. establishing general departmental policies and procedures.
B. evaluating department company officers during fire ground operations.
C. program scheduling.
D. All of the above

_____ **2.** An Instructor II responsibility is to ensure that the instructional team members realize that their primary role is to:

A. set organizational policies.
B. elicit feedback from learners and colleagues.
C. plan, develop, and conduct training sessions.
D. ensure that learner participation is encouraged.

_____ **4.** It is an Instructor II responsibility to ensure that the instructional team remains current. Professional development can be achieved by:

A. attending seminars.
B. attending workshops.
C. networking with other instructors.
D. All of the above

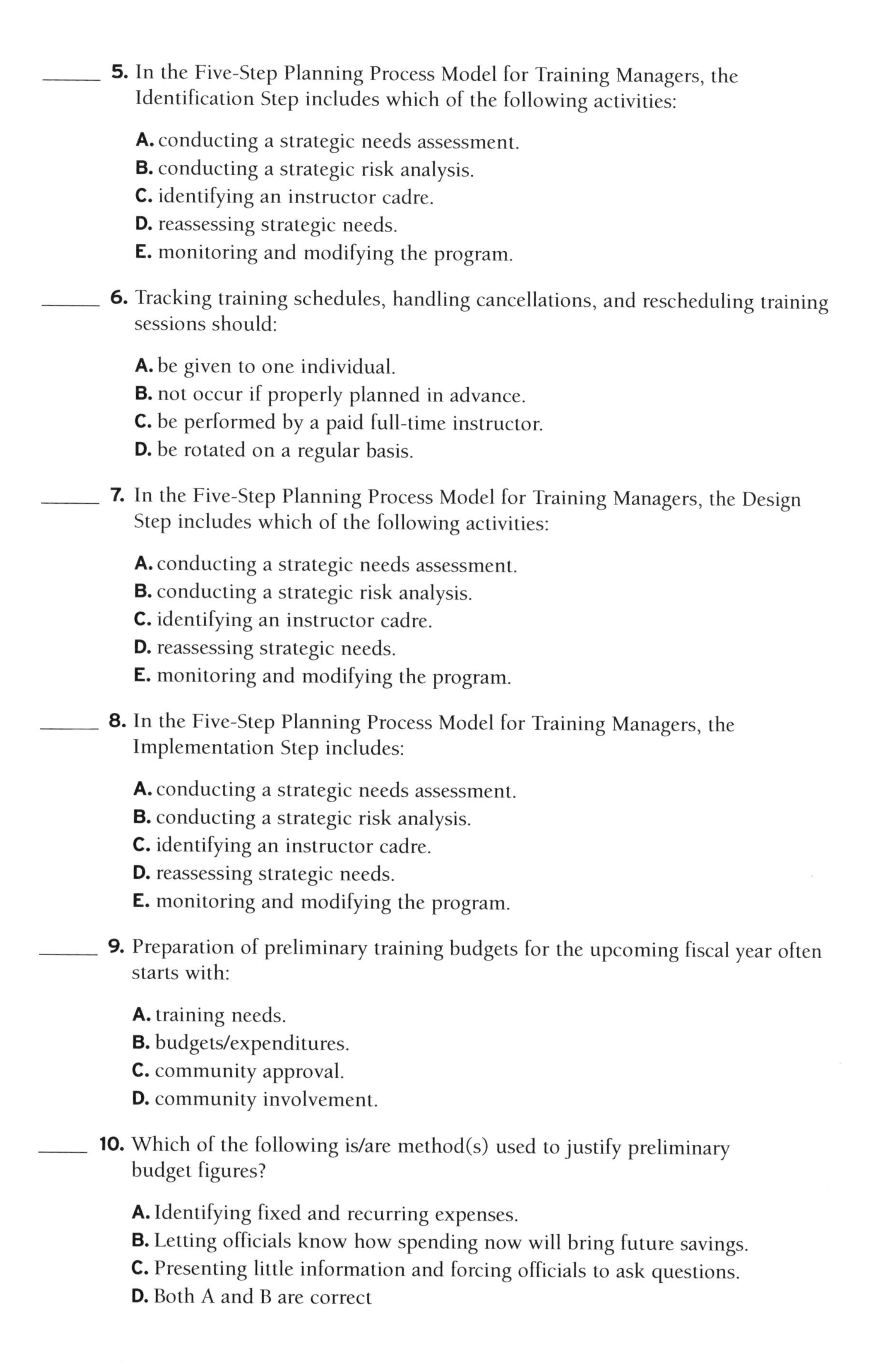

_____ **5.** In the Five-Step Planning Process Model for Training Managers, the Identification Step includes which of the following activities:

A. conducting a strategic needs assessment.
B. conducting a strategic risk analysis.
C. identifying an instructor cadre.
D. reassessing strategic needs.
E. monitoring and modifying the program.

_____ **6.** Tracking training schedules, handling cancellations, and rescheduling training sessions should:

A. be given to one individual.
B. not occur if properly planned in advance.
C. be performed by a paid full-time instructor.
D. be rotated on a regular basis.

_____ **7.** In the Five-Step Planning Process Model for Training Managers, the Design Step includes which of the following activities:

A. conducting a strategic needs assessment.
B. conducting a strategic risk analysis.
C. identifying an instructor cadre.
D. reassessing strategic needs.
E. monitoring and modifying the program.

_____ **8.** In the Five-Step Planning Process Model for Training Managers, the Implementation Step includes:

A. conducting a strategic needs assessment.
B. conducting a strategic risk analysis.
C. identifying an instructor cadre.
D. reassessing strategic needs.
E. monitoring and modifying the program.

_____ **9.** Preparation of preliminary training budgets for the upcoming fiscal year often starts with:

A. training needs.
B. budgets/expenditures.
C. community approval.
D. community involvement.

_____ **10.** Which of the following is/are method(s) used to justify preliminary budget figures?

A. Identifying fixed and recurring expenses.
B. Letting officials know how spending now will bring future savings.
C. Presenting little information and forcing officials to ask questions.
D. Both A and B are correct

____ **11.** The Instructor II must understand the five principles for budget management. Which of the following <u>**is not**</u> one of these principles?

A. Create/maintain records
B. Plan training around existing funding
C. Plan ahead
D. Share resources/create partnerships

____ **12.** Which item listed below is a capital expenditure?

A. Office supplies
B. Building
C. Heating and air conditioning
D. Insurance

____ **13.** Budget expenses categorized as human resources include:

A. worker's compensation.
B. office furniture.
C. office machines.
D. communication services.

____ **14.** Managing funding and resources to achieve training goals can be accomplished through which of the following nontraditional methods?

A. Developing cooperative relationships with the industry
B. Seeking funding through grants
C. Raising funds by providing training to the private sector
D. All of the above

____ **15.** According to the five principles for budget management, the Instructor II should:

A. base the training budget on program needs.
B. summarize actual expenditures and revenues.
C. determine what programs may be offered, depending on the size of the budget.
D. use alternative funds before using fixed budgetary funds.

____ **16.** ________________ is a broad, comprehensive term describing a person's or organization's responsibility under the law.

A. Guideline
B. Standard
C. Regulation
D. Liability

____ **17.** A policy is a guiding principle that organizations use to:

A. identify a general philosophy.
B. address specific issues or problems.
C. require a step-by-step outline of a task.
D. provide specific rules and regulations.

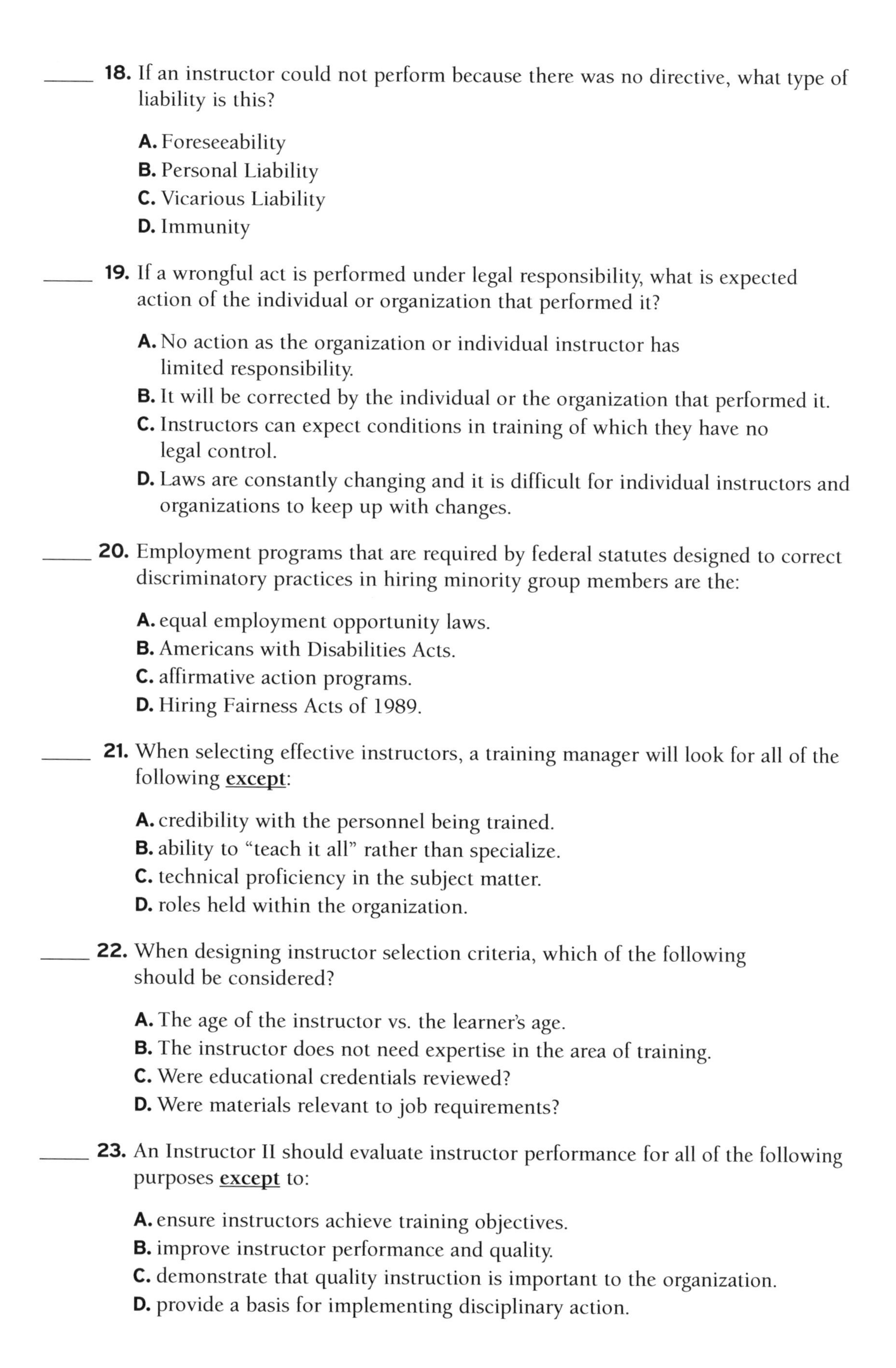

_____ **18.** If an instructor could not perform because there was no directive, what type of liability is this?

A. Foreseeability
B. Personal Liability
C. Vicarious Liability
D. Immunity

_____ **19.** If a wrongful act is performed under legal responsibility, what is expected action of the individual or organization that performed it?

A. No action as the organization or individual instructor has limited responsibility.
B. It will be corrected by the individual or the organization that performed it.
C. Instructors can expect conditions in training of which they have no legal control.
D. Laws are constantly changing and it is difficult for individual instructors and organizations to keep up with changes.

_____ **20.** Employment programs that are required by federal statutes designed to correct discriminatory practices in hiring minority group members are the:

A. equal employment opportunity laws.
B. Americans with Disabilities Acts.
C. affirmative action programs.
D. Hiring Fairness Acts of 1989.

_____ **21.** When selecting effective instructors, a training manager will look for all of the following **except**:

A. credibility with the personnel being trained.
B. ability to "teach it all" rather than specialize.
C. technical proficiency in the subject matter.
D. roles held within the organization.

_____ **22.** When designing instructor selection criteria, which of the following should be considered?

A. The age of the instructor vs. the learner's age.
B. The instructor does not need expertise in the area of training.
C. Were educational credentials reviewed?
D. Were materials relevant to job requirements?

_____ **23.** An Instructor II should evaluate instructor performance for all of the following purposes **except** to:

A. ensure instructors achieve training objectives.
B. improve instructor performance and quality.
C. demonstrate that quality instruction is important to the organization.
D. provide a basis for implementing disciplinary action.

_____ **24.** To effectively evaluate an instructor, the department should complete a form that:

A. identifies criteria for behaviors that are acceptable.
B. allows for multiple opinions to be given by different evaluators.
C. includes recommending disciplinary action.
D. includes opinions and identifies the names of learners participating in the evaluation.

_____ **25.** An Instructor II should perform instructor evaluations for the following purposes **except** to:

A. ensure that instructors achieve training objectives.
B. improve overall instructor performance and quality.
C. find ways to decrease the training budget.
D. demonstrate that quality instruction is important to the organization.

_____ **26.** As the evaluator in a Firefighter I class presentation, you observe the instructor using drawings, pictures, and transparencies. On your evaluation sheet, you would record that the lesson type was a/an:

A. illustration.
B. demonstration.
C. lecture.
D. simulation.

_____ **27.** When writing objectives, the _______________ describes how well the performance is to be accomplished.

A. course description
B. behavior
C. task analysis
D. standard

_____ **28.** The terms goals and objectives are often used interchangeably; however, these terms are not synonymous. Objectives:

A. identify teaching needs.
B. determine what needs are to be taught.
C. describe the end result.
D. give a time frame for completion.

_____ **29.** A manipulative skill lesson plan should include:

A. information sheets.
B. job breakdown sheets.
C. data sheets.
D. three levels of instruction.

_____ **30.** Given a length of 2 1/2 inch hose and a fog nozzle, the learner will connect the nozzle to the hose using the over-the-hip method **completing all steps with 100% accuracy within 10 seconds**. The underlined portion of the preceding statement represents the:

A. condition.
B. behavior.
C. standard.
D. application.

_____ **31.** A behavioral objective consists of the following components:

A. condition, situation, and behavior.
B. behavior, task, and condition.
C. behavior, standard, and time.
D. condition, criteria, and behavior.

_____ **32.** A behavioral objective **must** be stated:

A. at the beginning of each chapter.
B. clearly and distinctly by each learner.
C. in easy-to-understand terminology using a standard format.
D. in terms of measurable performance.

_____ **33.** In developing a job breakdown sheet, begin by listing _______________; then list the _______________ for performance that instructors must stress while teaching.

A. blocks; operations
B. operations; key points
C. operations; blocks
D. tasks; operations

_____ **34.** Which of the following **is not** part of an occupational analysis?

A. Unit
B. Block
C. Evaluation
D. Task

_____ **35.** On a screen, an instructor creates an image of a fire hazard situation that creates the illusion of fire and smoke. This statement **best** describes a/an:

A. simulation.
B. overhead projection.
C. multimedia presentation.
D. video tape.

_____ **36.** An **important** consideration when determining the need for a training aid is:

A. to support instruction and enhance learning.
B. whether it will be acceptable to the class.
C. the added time factor.
D. availability and cost factors.

_____ **37.** ________________ use a combination of audiovisual materials.

A. Easel pads
B. Illustrations
C. Cutaways
D. Multimedia

_____ **38.** When using nonprojected instructional media, advantages to using are that they are inexpensive, require little storage space, and provide a permanent record.

A. simulation aids
B. easel pads
C. dry easel boards
D. audio cassettes

_____ **39.** The task is a:

A. component of knowledge and skill in an occupation.
B. step which leads to another step.
C. description of what the learner is expected to do or the product or result.
D. career or professional category.

_____ **40.** There is **no purpose** in developing elaborate instructional material if:

A. enrollment is expected to be low.
B. budgeting constraints were not considered.
C. there are no instructors available to present the material.
D. there will be a great deal of resistance from the rank and file.

_____ **41.** Which of the following characterizes the affective domain of learning?

A. Evaluation
B. Adaptation
C. Valuing
D. Knowledge

_____ **42.** In the psychomotor domain of learning, the instructor must:

A. punish poor or unacceptable performance.
B. assure that each learner is allowed the same time period to comprehend the subject.
C. quickly move from one psychomotor level to another.
D. understand learner abilities at each level.

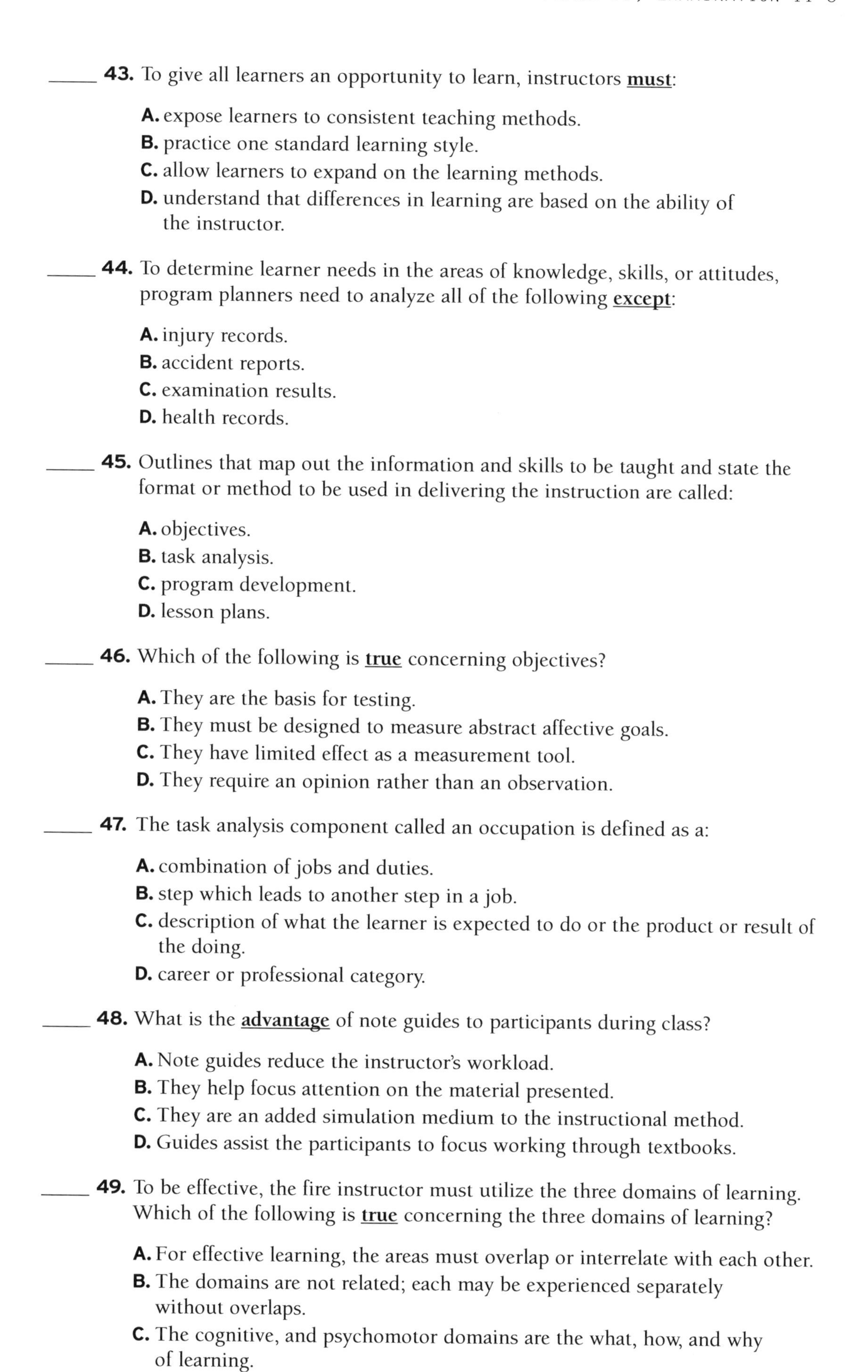

_____ **43.** To give all learners an opportunity to learn, instructors **must**:

A. expose learners to consistent teaching methods.
B. practice one standard learning style.
C. allow learners to expand on the learning methods.
D. understand that differences in learning are based on the ability of the instructor.

_____ **44.** To determine learner needs in the areas of knowledge, skills, or attitudes, program planners need to analyze all of the following **except**:

A. injury records.
B. accident reports.
C. examination results.
D. health records.

_____ **45.** Outlines that map out the information and skills to be taught and state the format or method to be used in delivering the instruction are called:

A. objectives.
B. task analysis.
C. program development.
D. lesson plans.

_____ **46.** Which of the following is **true** concerning objectives?

A. They are the basis for testing.
B. They must be designed to measure abstract affective goals.
C. They have limited effect as a measurement tool.
D. They require an opinion rather than an observation.

_____ **47.** The task analysis component called an occupation is defined as a:

A. combination of jobs and duties.
B. step which leads to another step in a job.
C. description of what the learner is expected to do or the product or result of the doing.
D. career or professional category.

_____ **48.** What is the **advantage** of note guides to participants during class?

A. Note guides reduce the instructor's workload.
B. They help focus attention on the material presented.
C. They are an added simulation medium to the instructional method.
D. Guides assist the participants to focus working through textbooks.

_____ **49.** To be effective, the fire instructor must utilize the three domains of learning. Which of the following is **true** concerning the three domains of learning?

A. For effective learning, the areas must overlap or interrelate with each other.
B. The domains are not related; each may be experienced separately without overlaps.
C. The cognitive, and psychomotor domains are the what, how, and why of learning.
D. All of the above

_____ **50.** Learning objectives:

A. indicate what instructors should teach.
B. are written for instructors.
C. provide instructors with verbal requirements.
D. describe what the learner will accomplish.

_____ **51.** Of the following components, which **is not** a major factor of learning?

A. Sensory memory
B. Short-term memory
C. Long-term memory
D. Cognitive memory

_____ **52.** Which of the following **is not** a component of a learning objective?

A. The audience
B. Criterion of performance
C. End behavior
D. Learner restrictions

_____ **53.** The principles of readiness, exercise, effect, disuse, association, recency, primacy, and intensity pertain to:

A. Maslow's Hierarchy of Needs.
B. Laws of Learning.
C. Herzberg's Theory.
D. Theory Y.

_____ **54.** There are _______________ steps of instruction.

A. one
B. two
C. three
D. four

_____ **55.** Certain assumptions can be made when teaching adults. These include:

A. adults need to be self-directed.
B. adults will learn what they need to know to meet job requirements.
C. adults' orientation is problem-centered.
D. All of the above

_____ **56.** A form used to provide the learner with opportunities to use steps or use multiple skills to complete an activity is the:

A. study sheet.
B. job breakdown sheet.
C. information sheet.
D. task worksheet.

_____ **57.** In psychomotor learning, the coaching process includes all of the following **except**:

A. observation.
B. evaluation.
C. suggestions.
D. discipline.

_____ **58.** Analyze the following comments and select the **most effective** coaching technique.

A. "That's wrong. Do it again, but do it the right way this time."
B. "Before you go any further, do you remember the next step?"
C. "You're not going to pass until we finish this lesson."
D. "Why can't you be more like my last class? They understood this topic."

_____ **59.** The discussion method should be used whenever possible if:

A. the instructor has limited time for delivery.
B. the instructor does not have time to prepare a lecture.
C. it precedes a simulation exercise.
D. the learners have sufficient knowledge of the subject.

_____ **60.** ________________ allow an instructor to evaluate the success of a presentation and emphasize important points.

A. Demonstrations
B. Questions
C. Conferences
D. Evolutions

_____ **61.** The method of instruction that requires the **most direct** instructor participation is:

A. role playing.
B. guided discussion.
C. simulation.
D. case study.

_____ **62.** One of the **most useful** teaching skills an instructor has is:

A. questioning techniques.
B. practical evaluations.
C. lectures.
D. written tests.

_____ **63.** Which method of instruction should be used in developing a learner's ability to analyze a situation and examine facts to reach a conclusion?

A. Role playing
B. Case study
C. Brainstorming
D. Discussion

_____ **64.** To conduct a successful conference, an instructor should:

A. allow spontaneous participation.
B. assist learners in realizing that their knowledge is limited.
C. provide clear direction toward a clearly stated end result.
D. emphasize the conference leader's knowledge of the subject.

_____ **65.** The ________________ reflects the final expected outcome of a lesson.

A. level of instruction
B. task to be accomplished
C. task content
D. self-satisfaction step

_____ **66.** Effective handling of a question includes all of the following **except**:

A. saying, "I don't know, but I'll find out."
B. answering the question thoroughly and in depth when the question is of interest to only the person asking.
C. referring the learner to related information when there is no exact answer to the question.
D. when a question refers to material that is covered in a later lesson, answering the question briefly and explaining that it will be covered in more detail later.

_____ **67.** If the weather is extremely hot, the instructor must:

A. give learners frequent breaks.
B. check administrative policy about training in these conditions.
C. consider safety factors first; have learners dressed in most comfortable and safe equipment.
D. All of the above

_____ **68.** During the ________________ step of the instructional delivery, the instructor should get the learners' attention, arouse curiosity, and develop interest.

A. lecture
B. preparation
C. application
D. demonstration

_____ **69.** When inclement weather is possible:

A. programs will not be affected. Training continues in all weather.
B. programs must have flexibility built in. Training can continue in some inclement weather.
C. programs should not be offered. Learning cannot take place if there is discomfort.
D. programs should be offered. Learners must become comfortable in working in these conditions.

_____ **70.** When teaching safety rules, regulations, and practices, instructors should make students aware that accident investigations are conducted to:

A. determine who is at fault.
B. determine why an accident occurred.
C. prevent a reoccurrence of the same or similar accidents.
D. Both B and C are correct.

_____ **71.** The first step to a safe training program is:

A. accident prevention.
B. environmental regulation.
C. proper supervision.
D. written rules and regulations.

_____ **72.** By regularly conducting an occupational or job analysis of a firefighter, the fire instructor is:

A. reducing vicarious liability.
B. practicing foreseeability.
C. justifying fireground injuries.
D. performing limited immunity.

_____ **73.** In preparation for high-hazard sessions, instructors must train and supervise learners in safety procedures, including:

A. oral guidelines only.
B. zero tolerance for disruptive behavior.
C. setting one set of standards for learners and another for instructors.
D. requesting learners to limit horseplay.

_____ **74.** NFPA 1403 sets the learner-to-teacher ratio for Live Fire training at:

A. 10 to 1.
B. 7 to 1.
C. 5 to 1.
D. 3 to 1.

_____ **75.** Acquired structures to be used for high-hazard training present difficult decisions for the instructor that include consideration of all of the following <u>**except**</u>:

A. environmental laws.
B. the designation of a structure as a historical landmark.
C. the structural safety of the structure/landmark.
D. the cost to the building owner.

_____ **76.** The instructor should ensure that water supply operations during high-hazard training, such as live burns:

A. are reliable for the entire duration of the exercise.
B. give students practical, on-the-job training.
C. are remote from the incident to allow practice in relay operations.
D. are performed by certified firefighters only.

_____ **77.** At a training evolution, the ________________ sector could refill air cylinders or provide a rehab area.

A. logistics
B. planning
C. operations
D. command

_____ **78.** Tanks used for training in confined-space rescue should:

A. expose the student to the same dangers expected in a real emergency.
B. have only one method for both access and egress.
C. contain simulated smoke to limit visibility.
D. have one side or end cut away to facilitate a safer training environment.

_____ **79.** A test used to measure an individual's proficiency in accomplishing a job or evolution is:

A. prescriptive.
B. progress.
C. multiple choice.
D. performance.

_____ **80.** A well-constructed ________________ test is generally recognized as one of the **most versatile** of the objective tests.

A. matching
B. multiple choice
C. true/false
D. essay

_____ **81.** ________________ tests are typically given in the middle or at the end of instruction.

A. Prescriptive
B. Application
C. Comprehensive
D. Both A and C are correct

_____ **82.** When preparing oral test items, the instructor should remember that they are difficult to score consistently and:

A. they are usually given in conjunction with a performance test.
B. people express the same ideas in different ways.
C. are usually administered in one-on-one situations.
D. All of the above

_____ **83.** A/an _________________ test requires learners to analyze, revise, redesign, or evaluate a problem.

A. objective
B. subjective
C. oral
D. performance

_____ **84.** True/false, multiple choice, and matching tests are examples of _________________ tests.

A. objective
B. subjective
C. oral
D. performance

_____ **85.** A written test that **minimizes** the possibility of learner guessing is the:

A. short answer.
B. true/false.
C. multiple choice.
D. matching.

_____ **86.** To identify learner mastery or non-mastery of subject matter, a _________________ test should be administered.

A. criterion-referenced
B. diagnosis
C. norm-referenced
D. classification

_____ **87.** A test designed to measure learner improvement throughout a course or unit of instruction is a _________________ test.

A. progress
B. summative
C. standardized
D. norm-referenced

_____ **88.** An **advantage** of matching tests is that they:

A. minimize guessing.
B. measure complete understanding.
C. easy to construct.
D. Both A and C are correct

_____ **89.** The correct choice in a multiple-choice test is referred to as the answer; the remaining choices are called:

A. negatives.
B. positives.
C. stems.
D. distractors.

_____ **90.** ________________ tests may be used to supplement performance tests to determine whether the learner knows the reasoning behind the jobs being performed.

A. Progress
B. Oral
C. Manipulative
D. Norm-referenced

_____ **91.** Adjusting or modifying examinations or training materials is an example of a reasonable accommodation consistent with the:

A. Americans with Disabilities Act.
B. Civil Rights Act.
C. Equal Employment Opportunity Laws.
D. Privacy Acts.

_____ **92.** A criterion reference test measures all of the following **except**:

A. the learner's performance in comparison to the program objectives.
B. the learner's level of mastery of the criterion requirements.
C. the learner's performance in comparison to the performance of others.
D. acceptability or unacceptability based on whether or not performance meets all conditions of criterion requirements.

_____ **93.** In evaluating a course, it is critical to make clear to anyone involved ________________ is to be evaluated and ________________.

A. what; why
B. who; when
C. what; where
D. who; how

_____ **94.** The **primary** purpose of course evaluation is to provide the instructor with feedback to improve the _______________ process.

A. teaching/learning
B. discussion
C. skill/demonstration
D. testing

_____ **95.** A final course evaluation feedback determines:

A. whether the instructional process has met the course objectives.
B. average learner scores.
C. learner participation in the course.
D. whether the course is needed or not.

_____ **96.** A _______________ evaluation is a post-course appraisal.

A. formative
B. practical
C. summative
D. All of the above

_____ **97.** Formative evaluation looks at the course during the _______________, while summative evaluation looks at the course after the _______________.

A. teaching; learning
B. delivery; program
C. product; procedure
D. teaching; process

_____ **98.** **Directions:** Read the statements below, then choose the correct answer.

1. Methods of analyzing test results are generally referred to as "test statistics."
2. Statistics are a way of organizing, analyzing, and interpreting test scores.
3. Analyzing tests has no affect in determining test validity.

A. Statement 1 is true; statement 2 is false; statement 3 is false.
B. Statement 1 is true; statement 2 is true; statement 3 is false.
C. Statement 1 is false; statement 2 is true; statement 3 is false.
D. Statement 1 is false; statement 2 is false; statement 3 is true.

_____ **99.** Instructors who analyze student feedback may determine:

A. that students are biased and unable to recognize the instructor's depth of knowledge.
B. areas to adjust and improve.
C. low scores mean that the teacher did not teach.
D. that students are unable to be objective.

____ **100.** In test analysis, raw score consists of:

A. points learner receives on a test.
B. percentage score times median score.
C. median score plus the mean.
D. percentage score added to the mean.

Did you score higher than 80% on Examination II-3? Circle Yes or No in ink. (We will return to your answer to this question later in SAEP.)

Feedback Step

Now, what do we do with your "yes" and "no" answers given throughout the NFPA Standard 1041 examination preparation process? First, return to any response that has "no" circled. Go back to the highlighted answers for those examination items missed. Read and study the paragraph preceding the location of the answer, as well as the paragraph following the paragraph where the answer is located. This will expand your knowledge base for the missed question, put it in a broader perspective, and improve associative learning. Remember, you are trying to develop mastery of the required knowledge. Scoring 80 percent on an examination is good, but it is not mastery performance. To be at the top of your group, you must score much more than 80 percent on your training, promotion, or certification examination.

Phases III and IV focus on getting you ready for the examination process by recommending activities that have a positive impact on the emotional and the physical aspects of examination preparation. By evaluating your own progress through SAEP, you have determined that you have a high level of knowledge. Taking an examination for training, promotion, or certification is a competitive event. Just as in sports, total preparation is vitally important. Now you need to get all the elements of good preparation in place so that your next examination experience will be your best ever. Phase III is next!

Carefully review the Summary of Key Rules for Taking an Examination and Summary of Helpful Hints. Do this review now and at least two additional times prior to taking your next examination.

Summary of Key Rules for Taking an Examination

Rule 1—Examination preparation is not easy. Preparation is 95% perspiration and 5% inspiration.

Rule 2—Follow the steps very carefully. Do not try to reinvent or shortcut the system. It really works just as it was designed to!

Rule 3—Mark with an "X" any examination items for which you guessed the answer. For maximum return on effort, you should also research any answer that you guessed, even if you guessed correctly. Find the correct answer, highlight it, and then read the entire paragraph that contains the answer. Be honest and mark all questions you guessed. Some examinations have a correction for guessing built into the scoring process. The correction for guessing can reduce your final examination score. If you are guessing, you are not mastering the material.

Rule 4—Read questions twice if you have any misunderstanding and especially if the question contains complex directions or activities.

Rule 5—If you want someone to perform effectively and efficiently on the job, the training and testing program must be aligned to achieve this result.

Rule 6—When preparing examination items for job-specific requirements, the writer must be a subject matter expert with current experience at the level that the technical information is applied.

Rule 7— Good luck = good preparation.

Summary of Helpful Hints

Helpful Hint - Most of the time your first impression is the best. More than 41% of changed answers during PTS's SAEP field test were changed from a right answer to a wrong answer. Another 33% were changed from a wrong answer to another wrong answer. Only 26% of answers were changed from wrong to right. In fact three participants did not make a perfect score of 100% because they changed one right answer to a wrong one! Think twice before you change your answer. The odds are not in your favor.

Helpful Hint - Researching correct answers is one of the most important activities in SAEP. Locate the correct answer for all missed examination items. Highlight the correct answer. Then read the entire paragraph containing the answer. This will put the answer in context for you and provide important learning by association.

Helpful Hint - Proceed through all missed examination items using the same technique. Reading the entire paragraph improves retention of the information and helps you develop an association with the material and learn the correct answers. This step may sound simple. A major finding during the development and field testing of SAEP was that you learn from your mistakes.

Helpful Hint - Follow the steps carefully to realize the best return on effort. Would you consider investing your money in a venture without some chance of earning a return on that investment? Examination preparation is no different. You are investing time and expecting a significant return for that time. If, indeed, time is money, then you are investing money and are due a return on that investment. Doing things right and doing the right things in examination preparation will ensure the maximum return on effort.

Helpful Hint - Try to determine why you selected the wrong answer. Usually something influenced your selection. Focus on the difference between your wrong answer and the correct answer. Carefully read and study the entire paragraph containing the correct answer. Highlight the answer just as you did for the other examinations.

Helpful Hint - Studying the correct answers for missed items is a critical step in achieving your desired return on effort! The focus of attention is broadened, and new knowledge is often gained by expanding association and contextual learning. During PTS's research and field test, self-study during this step of SAEP resulted in gains of 17 points between the first examination administered and the third examination. An increase in your score of 17 points can move you from the lower middle to the top of the list of persons taking a training, promotion, or certification examination. That is a competitive edge and a prime example of return on effort in action. Remember: Maximum effort = maximum results!

PHASE III

How Examination Developers Think – Getting Inside Their Heads

Now that you have finished the examination practice, this additional information will assist you in understanding and applying examination-taking skills. Developing your knowledge of how examination professionals think and prepare examinations is not cheating. Most serious examination takers have spent many hours reviewing various examinations to gain an insight into the technology used to develop them. It is a demanding technology when used properly. You probably already know this if you have prepared examination items and administered them in your fire department.

Phase III will not cover all the ways and means of examination-item writing. Examination-item writers use far too many techniques to cover adequatly in this book. Instead, the focus here is on key techniques that will help you achieve a better score on your examination.

How are examination items derived?

Professional examination-item writers use three basic techniques to derive examination items from text or technical reference materials: verbatim, deduction, and inference.

The most common technique is to take examination items verbatim from materials in the reference list. This technique doesn't work well for mastering information, however. The verbatim form of testing encourages rote learning-that is, simply memorizing the material. The results of this type of learning are not long-lasting, nor are they appropriate for learning and retaining the critical knowledge that you must have for on-the-job performance. Consequently, SAEP doesn't create the majority of examination questions covering NFPA Standard 1041 using the verbatim technique.

Professional examination-item writers tend to use verbatim testing at the most basic level of job classifications. A first responder, for instance, is expected to learn many basic facts. At this level, verbatim examination items can be justified.

In the higher ranks of the Fire and Emergency Medical Service, other types of examination items are more beneficial and productive for mastering higher cognitive knowledge and skills. At the higher cognitive levels of an occupation, such as Fire Officer, examination development will therefore rely on other means. The most important technique at the higher cognitive levels is using deduction as the basis for examination items. This technique exercises logic and analytical skills and often requires the examination taker to read materials several times to answer the examination item. It is not, then, a matter of simply repeating the information that results in a verbatim answer.

At the first responder level, most activities are carefully supervised by a more experienced technician or company officer. At this level, the responder is expected to closely follow commands and is encouraged not to use deductive reasoning that can lead to "freelance" responder tactics. As one progresses to the Fire Instructor II level and gains experience, however, deductive reasoning and inferences skills are developed and applied. Most of these skills are related to personal safety and the safety of those on the scene. Most size up and strategies are developed and passed from the officers on the scene to the first responders.

Rule 5

If you want someone to perform effectively and efficiently on the job, the training and testing program must be aligned to achieve this result.

Rule #5 is of paramount importance for first responders. Effective and efficient first responders are able to receive fireground commands, follow instructions, and perform their tasks as safely and as rapidly as they can. There are limited opportunities for first responders to do much else, because they serve as the first line of action at the emergency scene.

Consider the following example of deductive reasoning: An incident call is received from the telecommunicator stating that an infant has a high temperature and is convulsing. Just this amount of information should cause the first responder to immediately plan the response, conduct size-up activities, and review infant care procedures en route. Some of these deductive responses will have you focus on the infant's age, past medical history, location, access, and many other possible factors. If you have an EMT or paramedic background, a list of several items could be deduced that would expedite an efficient and effective response to the incident.

You can probably think of many first responder tasks and circumstances that rely on deductive reasoning. The more experience you gain on the fireground as a fire fighter, the more often you will be called upon to practice deductive reasoning and inference from emergency data, and the more efficient and effective you will become, whether the situation involves ventilating a roof or attending to the emergency needs of an infant.

Legendary football coach Vince Lombardi was once asked about the precision performance of his offensive and defensive teams. It was suggested that Lombardi must spend a great deal of time on the practice field to achieve those results. Lombardi responded, "Practice doesn't make perfect; only perfect practice makes perfect." This is exactly what is required to be an outstanding examination taker. Most people don't perfectly practice examination-taking skills.

A third technique used by professional examination-item writers is to rely on inference or implied answers to develop examination items. Inference requires contrasting, comparing, analyzing, evaluating, and other high-level cognitive skills. Tables, charts, graphs, and other instruments for presenting data provide excellent means for deriving inference-based examination items. Implied answers are based on logic. They rely on your ability to use logical processes or series of facts to arrive at a plausible answer.

For example, recent data gathered by the NFPA stated that heart attacks remain the leading cause of death for fire service personnel. Other NFPA-supplied data indicated that strains and sprains are the leading cause of injuries on the job. Several inferences can be made from these relatively simple statements. A safety officer can infer the results of the NFPA study to his or her own personnel and use the information as a trigger for checking on personnel, conducting surveys, reviewing accident records, and comparing the study results with actual experience. Is that particular fire department doing better or worse in terms of these important health issues? Are the Fire and Emergency Medical Service personnel getting the right exercise? Are they diligent in keeping the station and fireground free from the activities that may lead to strains and sprains? The basic inference here is that any particular fire department may be similar or different in some ways from the generalized data.

Sometimes it may be difficult to find an answer to an examination item because it is measuring your ability to make deductions and draw inferences from the technical materials.

How are examination items written and validated?

Once the pertinent information is identified and the technique for writing an examination item selected, the professional examination-item writer will prepare a draft. The draft examination item is then referenced to specific technical information, such as a textbook, manufacturer's manual, or other related technical information. If the information is derived from a job-based requirement, then it should also be validated by job incumbents (i.e., those who are actually performing in the occupation at the specific level of the required knowledge).

Rule 6

When preparing examination items for job-specific requirements, the writer must be a subject matter expert with current experience at the level where the technical information is applied.

Rule #6 ensures that the examination item has a basic level of job content validity. The final level of job content validity is determined by using committees or surveys of job incumbents who certify the information to be current and required on the job. The information must be in a category of "need to know" or "must know" to be considered job relevant. The technical information must be accurate. Because subject matter experts do need basic training in examination-item writing, it is recommended that a professional in examination technology be part of the review process so that basic rules and guidelines of the industry are followed.

Finally, the examination items must be field tested. Once this testing is complete, statistical and analytical tools are available to help revise and improve the examination items. These techniques and tools go well beyond the scope of this Exam Prep book. Professionals are available to conduct these data analyses, and their services should be used.

Good Practices in Examination Item and Examination Development

The most reliable examinations are objective. That is, each question has only one answer that is accepted by members of the occupation. This objective quality permits fair and equitable examinations. The most popular types of objective examination items are multiple choice, true/false, matching, and completion (fill in the blanks).

Valid and reliable job-relevant examinations for the Fire and Emergency Medical Service industry must satisfy 10 rules:

1. They do not contain trick questions.
2. They are short and easy to read, using language and terms appropriate to the target examination population.
3. They are supported with technical references, validation information, and data on their difficulty, discrimination, and other item analysis statistics.
4. They are formatted to meet recognized testing standards and examples.
5. They focus on the "need to know" and "must know" aspects of the job.
6. They are fair and objective.
7. They are not based on obscure and trivial knowledge and skills.
8. They can be easily defended in terms of job-content requirements.
9. They meet national and other professional job qualification standards.
10. They demonstrate their usefulness as part of a comprehensive testing program, including written, oral, and performance examination items.

The primary challenges of job-relevant examinations relate to their currency and validity. Careful recording of data, technical reference sources, and the examination writer's qualifications are important. Examinations that affect someone's ability to be promoted, certified, or licensed, as well as to complete training that leads to a job, have exacting requirements both in published documents and in the laws of the land.

Three Common Myths of Examination Construction

1. **Myth:** If in doubt about the answer for a multiple-choice examination item, select the longest answer.

   Reality: Professional examination-item writers use short answers as correct ones at an equal or higher percentage than longer answers. Remember, there are usually choices A-D. That leaves three other possibilities for the correct answer other than the longest one. Statistically speaking, the longest answer is less likely to be correct.

2. **Myth:** If in doubt about the answer for a multiple-choice examination item, select "C".

   Reality: Computer technology and examination-item banking permit multiple versions of examinations to be developed simultaneously. This is typically achieved by moving the correct answer to different locations (for example, version 1 will have the correct answer in the "C" position, version 2 will have it in the "D" position, and so forth).

3. **Myth:** Watch for errors in singular examination-item stems with plural choices in the A-D answers, or vice versa.

   Reality: Most computer-based programs have spelling and grammar checking utilities. If this mistake occurs, an editing error is the probable cause and usually has nothing to do with detecting the right answer.

Some Things That Work

1. Two to three days before your examination, review the examination items you missed in SAEP. Read those highlighted answers and the entire paragraph one more time.
2. During the examination, carefully read each examination item twice. Once you have selected your answer, read the examination item and answer together. This technique can prompt you to recall information that you studied during your examination preparation activities.
3. Apply what you learned in SAEP. Eliminate as many distracters as possible to improve your chance of answering the question correctly.
4. Pace yourself. Know how much time you have to take the examination. If an examination item is requiring too much time, write its number down and continue with the next examination item. Often, a later examination item will trigger your memory and make the earlier examination item seem easier to answer. (For a time pacing strategy, see the Examination Pacing Table at the end of Phase IV.)

5. Don't panic if you don't know some examination items. Leave them to answer later. The most important thing is to finish the examination, because there may be several examination items at the end of the examination that you do know.
6. As time runs out for taking the examination, do not panic. Concentrate on answering those difficult examination items that you skipped.
7. Double-check your answer sheets to make sure you have not accidentally left an answer blank.
8. Once you complete the examination, return to the difficult examination items. Often, while taking an examination, other examination items will cause you to remember or associate those answers with the difficult examination-item answers. The longer the examination, the more likely you will be to gather the information needed to answer more difficult examination items.

There are many other helpful hints that can be used to improve your examination-taking skills. If you want to research the materials on how to take examinations and raise your final score, visit your local library, a bookstore, or the Web for additional resources. The main reason we developed SAEP is to provide practice and help you develop examination-taking skills that you can use throughout your life.

PHASE IV

The Basics of Mental and Physical Preparation

Mental Preparation – I Can Get My Head Ready!

The two most common mental blocks to examination taking are examination anxiety and fear of failure. In the Fire and Emergency Medical Service, these feelings can create significant performance barriers. Overcoming severe conditions may require some professional psychological assistance, which is beyond the scope of this *Exam Prep* book.

The root cause of examination anxiety and fear of failure is often lack of self-confidence. SAEP was designed to help improve your self-confidence by providing evidence of your mastery of the material on the examination. Look at your scores as you progress through Phase I or Phase II. Review your Personal Progress Plotter; it will help you gain confidence in your knowledge of NFPA Standard 1041. Look at your Personal Progress Plotter the day before your scheduled examination and experience renewed confidence.

Let's examine the meaning of anxiety. Knowing what it is will help you deal with it at examination time. According to *Webster's Dictionary*, anxiety is "uneasiness and distress about future uncertainties." Many of us have real anxiety about taking examinations, and it is a natural response for some, often prefaced by questions like these: Am I ready for this? Do I have a good idea of what will be on the examination? Will I make the lowest score? Will John Doe score higher than me?

These questions and concerns are normal. Remember that hundreds of people have gone through SAEP and achieved an average gain of 17 points in their scores. The preparation process will help you maintain your self-confidence. Once again, review the evidence in your Personal Progress Plotter to see what you have accomplished.

Fear, according to *Webster's Dictionary*, is "alarm and agitation caused by the expectation or realization of danger." It is a normal reaction to examinations. To deal with it, first analyze the degree of fear you may be experiencing several days before the examination date. Then focus on the positive experiences you had as you finished SAEP. Putting your fear in perspective by using positives to eliminate or minimize it is a very important examination-taking skill. The more you focus on your positive accomplishments in mastering the materials, the less fear you will experience.

If your fear and anxiety persist even after you take steps to build your confidence, you may want to get some professional assistance. Do it now! Don't wait until the week before the examination. There may be real issues that a professional can help you deal with to overcome these feelings. Hypnosis and other forms of treatment have been found to be very helpful. Consult with an expert in this area.

Physical Preparation - Am I Really Ready?

Physical preparation is the element that is probably most ignored in examination preparation. In the Fire and Emergency Medical Service, examinations are often given at locations away from home. If this is the case, you need to be especially careful of key physical concerns. More will be said about that later.

In general, following these helpful hints will help you concentrate, enhance your examination performance, and add points to your score.

1. Do not "cram" for the examination. This factor was found to be first in importance during PTS's field test of SAEP. Cramming results in examination anxiety, adds to confusion, and tends to lessen the effectiveness of the examination-taking skills you already possess. Avoid cramming!
2. Get a normal night's rest. It may even be wise to take a day off before the examination to rest. Do not schedule an all-night shift right before your examination.
3. Avoid taking excessive stimulants or medications that inhibit your thinking processes. Eat at least three well-balanced meals before the day of the examination. It is a good practice to carry a balanced energy bar (not candy) and a bottle of water into the examination area. Examination anxiety and fear can cause a dry mouth, which can lead to further aggravation. Nibbling on the energy bar also has a settling effect and supplies some "brain food."
4. If the examination is taking place at an out-of-town location, do the following:

   - Avoid a "night out with friends." Lack of rest, partying, and fatigue are major examination performance killers.
   - Check your room carefully. Eliminate things that may aggravate you, interfere with your rest, or cause any discomfort. If the mattress isn't good, the pillows are horrible, or the room has an unpleasant odor, change rooms or even hotels.
   - Wake up in plenty of time to take a relaxing shower or soaking bath. Don't put yourself in a "rush" mode. Things should be carefully planned so that you arrive at the examination site ahead of time, calm, and collected.
5. Listen to the examination proctor. The proctor usually has rules that you must follow. Important instructions and directions are usually given. Ask clarifying questions immediately and listen to the responses to questions raised by the other examination takers. Most examination environments are carefully controlled and may not permit questions you raise that are covered in the proctor's comments or deal with the technical content in the examination itself. Be attentive, focus, and succeed.
6. Remain calm and breathe. Pace yourself. Apply your examination-taking skills learned during SAEP.
7. Remember the analogy of an examination as a competitive event. If you want to gain a competitive edge, carefully follow all phases of SAEP. This process has yielded outstanding results in the past and will do so for you.

Time Management During Examinations

The following table will help you pace yourself during an examination. You should become familiar with the table and be able to construct your own when you are in the examination room and getting ready to start the examination process. This effort will take a few minutes, but it will make a tremendous contribution to your time management during the examination.

Here is how the table works. First divide the examination time into six equal parts. For example, if you have 3 ½ hours (210 minutes) for the examination, then each of the six time parts contains 35 minutes (210 ÷ 6 = 35 minutes). Now divide the number of examination items by 5. For example, if the examination has 150 examination items, 150 ÷ 5 = 30. Now, with the math done, you can set up a table that tells you approximately how many examination items you should answer in 35 minutes (the equal time divisions).

You should be on or near examination item 30 at the end of the first 35 minutes, and so forth. Notice that we divided the number of examination items by 5 and the time available by 6. The extra time block of 35 minutes is used to double-check your answer sheet, focus on difficult questions, and calm your nerves. This technique will work wonders for your stress level and, yes, it will improve your examination score.

Examination Pacing Table (150 and 100 Examination Items)

Time for Examination	Minutes for Six Equal Time Parts	Number of Examination Items	Examination Items per Time Part	Time for Examination Review
210 minutes (3.5 hours)	35	150	30 (Number of examination items to be answered)	35 minutes (Chilling and double-checking examination)
150 minutes (2.5 hours)	25	100	20 (Number of examination items to be answered)	25 minutes (Chilling and double-checking examination)

The Examination Pacing Table can be altered by adjusting the time and examination item variables, as either may change in the real examination environment. For instance, if the time changes, adjust the amount of time available to answer the examination items in each of the five time blocks. If the examination item numbers increase or decrease, adjust the number of examination items to be answered in the time blocks.

Take some precautions when using this time management strategy:

1. Do not panic if you run a few minutes behind in each time block. This time management strategy should not stress you while you are using it. Most people tend to pick up their pace as they move into the examination.
2. During the examination, carefully mark or note examination items that you need to return to during your review time block. This will help you expedite your examination completion check.
3. Do not be afraid to ask for more time to complete your examination. In most cases, the time limit is flexible or should be.
4. Double-check your answer sheet to make sure that you didn't leave blank responses and that you didn't double-mark answers. Double-markings are most often counted as wrong answers. Make sure that any erasures are made cleanly. Caution: When you change your answer, make sure that you really want to do so. The odds are not in your favor unless something on the examination really influenced the change.

APPENDIX A

Examination I-1 Answer Key

Directions

Follow these steps carefully for completing the feedback part of SAEP:

1. After calculating your score, look up the answers for the examination items you missed as well as those on which you guessed, even if you guessed correctly. If you are guessing, it means the answer is not perfectly clear. In this process, we are committed to making you as knowledgeable as possible.
2. Enter the number of missed and guessed examination items in the blanks on your Personal Progress Plotter.
3. Highlight the answer in the reference materials. Read the paragraph preceding and the paragraph following the one in which the correct answer is located. Enter the paragraph number and page number next to the guessed or missed examination item on your examination. Count any part of a paragraph at the beginning of the page as one paragraph until you reach the paragraph containing your highlighted answer. This step will help you locate and review your missed and guessed examination items later in the process. It is essential to learning the material in context and by association. These learning techniques (context/association) are the very backbone of the SAEP approach.
4. Once you have completed the feedback part, you may proceed to the next examination.

1. Reference: NFPA 1041, 4.1.1
IFSTA, *Fire and Emergency Services Instructor,* 6th Edition, 4th Printing, page vii.
Answer: B

2. Reference: NFPA 1041, 4.1.1, 4.3.1, and 4.4.1
IFSTA, *Fire and Emergency Services Instructor,* 6th Edition, 4th Printing, pages 3–4.
Answer: A

3. Reference: NFPA 1041, 4.2.3 and 4.2.3(a)(b)
IFSTA, *Fire and Emergency Services Instructor,* 6th Edition, 4th Printing, page 252.
Answer: C

4. Reference: NFPA 1041, 4.2.3 and 4.2.3(a)(b)
IFSTA, *Fire and Emergency Services Instructor,* 6th Edition, 4th Printing, page 252.
Answer: C

5. Reference: NFPA 1041, 4.2.3, 4.2.3(a)(b), 4.4.2, and 4.4.2(a)(b)
IFSTA, *Fire and Emergency Services Instructor,* 6th Edition, 4th Printing, page 28.
Answer: C

6. Reference: NFPA 1041, 4.2.3 and 4.2.3(a)(b)
IFSTA, *Fire and Emergency Services Instructor,* 6th Edition, 4th Printing, page 29.
Answer: D

7. Reference: NFPA 1041, 4.3.1 and 3.3.2.1
IFSTA, *Fire and Emergency Services Instructor,* 6th Edition, 4th Printing, page 4.
Answer: C

8. Reference: NFPA 1041, 4.3.2, 4.3.2(a)(b), and 4.3.1
IFSTA, *Fire and Emergency Services Instructor,* 6th Edition, 4th Printing, page 98.
Answer: A

9. Reference: NFPA 1041, 4.3.2 and 4.3.2(a)
IFSTA, *Fire and Emergency Services Instructor,* 6th Edition, 4th Printing, pages 68–69.
Answer: D

10. Reference: NFPA 1041, 4.3.2 and 4.3.2(a)(b)
IFSTA, *Fire and Emergency Services Instructor,* 6th Edition, 4th Printing, page 133.
Answer: C

11. Reference: NFPA 1041, 4.3.2 and 4.3.2(a)(b)
IFSTA, *Fire and Emergency Services Instructor,* 6th Edition, 4th Printing, page 222.
Answer: B

12. Reference: NFPA 1041, 4.3.2 and 4.3.2(a)(b)
IFSTA, *Fire and Emergency Services Instructor,* 6th Edition, 4th Printing, page 96.
Answer: C

13. Reference: NFPA 1041, 4.3.2, 4.3.2(a)(b), 4.3.3, and 4.3.3(a)(b)
IFSTA, *Fire and Emergency Services Instructor,* 6th Edition, 4th Printing, page 96.
Answer: D

14. Reference: NFPA 1041, 4.3.3 and 4.3.3(a)(b)
IFSTA, *Fire and Emergency Services Instructor,* 6th Edition, 4th Printing, pages 117–118.
Answer: C

15. Reference: NFPA 1041, 4.3.3 and 4.3.3(a)(b)
IFSTA, *Fire and Emergency Services Instructor,* 6th Edition, 4th Printing, pages 96 and 390.
Answer: C

16. Reference: NFPA 1041, 4.3.3 and 4.3.3(a)(b)
IFSTA, *Fire and Emergency Services Instructor,* 6th Edition, 4th Printing, pages 58-59 and 389.
Answer: B

17. Reference: NFPA 1041, 4.3.3 and 4.3.3(a)(b)
IFSTA, *Fire and Emergency Services Instructor,* 6th Edition, 4th Printing, page 114.
Answer: D

18. Reference: NFPA 1041, 4.3.3 and 4.3.3(a)(b)
IFSTA, *Fire and Emergency Services Instructor,* 6th Edition, 4th Printing, page 6.
Answer: A

19. Reference: NFPA 1041, 4.3.3 and 4.3.3(a)(b)
IFSTA, *Fire and Emergency Services Instructor,* 6th Edition, 4th Printing, page 120.
Answer: D

20. Reference: NFPA 1041, 4.3.3 and 4.3.3(a)(b)
IFSTA, *Fire and Emergency Services Instructor,* 6th Edition, 4th Printing, page 96.
Answer: A

21. Reference: NFPA 1041, 4.3.3, 4.3.3(a)(b), 4.4.3, and 4.4.3(a)(b)
IFSTA, *Fire and Emergency Services Instructor,* 6th Edition, 4th Printing, page 7.
Answer: A

22. Reference: NFPA 1041, 4.3.3 and 4.3.3(a)(b)
IFSTA, *Fire and Emergency Services Instructor,* 6th Edition, 4th Printing, pages 104 and 390.
Answer: C

23. Reference: NFPA 1041, 4.3.3 and 4.3.3(a)(b)
IFSTA, *Fire and Emergency Services Instructor,* 6th Edition, 4th Printing, page 96.
Answer: A

24. Reference: NFPA 1041, 4.4.2, 4.4.2(a)(b), 4.4.3, and 4.4.3(a)(b)
IFSTA, *Fire and Emergency Services Instructor,* 6th Edition, 4th Printing, page 93.
Answer: C

25. Reference: NFPA 1041, 4.4.2, 4.4.2(a)(b), 4.4.5, and 4.4.5(a)(b)
IFSTA, *Fire and Emergency Services Instructor,* 6th Edition, 4th Printing, page 29.
Answer: A

26. Reference: NFPA 1041, 4.4.2, 4.4.2(a)(b) 4.4.3, and 4.4.3(a)(b)
IFSTA, *Fire and Emergency Services Instructor,* 6th Edition, 4th Printing, page 118.
Answer: C

27. Reference: NFPA 1041, 4.4.2 and 4.4.2(a)(b)
IFSTA, *Fire and Emergency Services Instructor,* 6th Edition, 4th Printing, page 149.
Answer: C

28. Reference: NFPA 1041, 4.4.2 and 4.4.2(a)(b)
IFSTA, *Fire and Emergency Services Instructor,* 6th Edition, 4th Printing, pages 24 and 146.
Answer: C

29. Reference: NFPA 1041, 4.4.2 and 4.4.2(a)
IFSTA, *Fire and Emergency Services Instructor,* 6th Edition, 4th Printing, page 23.
Answer: A

30. Reference: NFPA 1041, 4.4.2 and 4.4.2(a)(b)
IFSTA, *Fire and Emergency Services Instructor,* 6th Edition, 4th Printing, pages 148–149
Answer: C

31. Reference: NFPA 1041, 4.4.2 and 4.4.2(a)(b)
IFSTA, *Fire and Emergency Services Instructor,* 6th Edition, 4th Printing, page 164.
Answer: A

32. Reference: NFPA 1041, 4.4.3 and 4.4.3(a)
IFSTA, *Fire and Emergency Services Instructor,* 6th Edition, 4th Printing, pages 72-73.
Answer: B

33. Reference: NFPA 1041, 4.4.3 and 4.4.3(a)(b)
IFSTA, *Fire and Emergency Services Instructor,* 6th Edition, 4th Printing, page 158.
Answer: C

34. Reference: NFPA 1041, 4.4.3 and 4.4.3(a)(b)
IFSTA, *Fire and Emergency Services Instructor,* 6th Edition, 4th Printing, page 7.
Answer: B

35. Reference: NFPA 1041, 4.4.3, 4.4.3(a)(b), 4.4.4, and 4.4.4(a)
IFSTA, *Fire and Emergency Services Instructor,* 6th Edition, 4th Printing, page 166.
Answer: D

36. Reference: NFPA 1041, 4.4.3 and 4.4.3(a)
IFSTA, *Fire and Emergency Services Instructor,* 6th Edition, 4th Printing, page 74.
Answer: B

37. Reference: NFPA 1041, 4.4.3, 4.4.3(a)(b), 4.4.5, and 4.4.5(a)(b)
IFSTA, *Fire and Emergency Services Instructor,* 6th Edition, 4th Printing, pages 64–65.
Answer: D

38. Reference: NFPA 1041, 4.4.3, 4.4.3(a)(b), 4.4.5, and 4.4.5(a)(b)
IFSTA, *Fire and Emergency Services Instructor,* 6th Edition, 4th Printing, pages 68–69.
Answer: D

39. Reference: NFPA 1041, 4.4.3 and 4.4.3(a)
IFSTA, *Fire and Emergency Services Instructor,* 6th Edition, 4th Printing, page 74.
Answer: A

40. Reference: NFPA 1041, 4.4.3, 4.4.3(a)(b), 4.4.5, and 4.4.5(a)(b)
IFSTA, *Fire and Emergency Services Instructor,* 6th Edition, 4th Printing, page 65.
Answer: C

41. Reference: NFPA 1041, 4.4.3 and 4.4.3(a)(b)
IFSTA, *Fire and Emergency Services Instructor,* 6th Edition, 4th Printing, page 161.
Answer: C

42. Reference: NFPA 1041, 4.4.3, 4.4.3(a)(b), 4.4.4, and 4.4.4(a)
IFSTA, *Fire and Emergency Services Instructor,* 6th Edition, 4th Printing, pages 56–57.
Answer: C

43. Reference: NFPA 1041, 4.4.3, 4.4.3(a), 4.4.5, and 4.4.5(a)
IFSTA, *Fire and Emergency Services Instructor,* 6th Edition, 4th Printing, page 54.
Answer: C

44. Reference: NFPA 1041, 4.4.3, 4.4.3(a)(b), and 1.3.5
IFSTA, *Fire and Emergency Services Instructor,* 6th Edition, 4th Printing, page 180.
Answer: C

45. Reference: NFPA 1041, 4.4.3 and 4.4.3(a)(b)
IFSTA, *Fire and Emergency Services Instructor,* 6th Edition, 4th Printing, page 8.
Answer: B

46. Reference: NFPA 1041, 4.4.4 and 4.4.4(a)
IFSTA, *Fire and Emergency Services Instructor,* 6th Edition, 4th Printing, page 9.
Answer: D

47. Reference: NFPA 1041, 4.4.4 and 4.4.4(a)
IFSTA, *Fire and Emergency Services Instructor,* 6th Edition, 4th Printing, page 144.
Answer: C

48. Reference: NFPA 1041, 4.4.5 and 4.4.5(a)(b)
IFSTA, *Fire and Emergency Services Instructor,* 6th Edition, 4th Printing, page 136.
Answer: D

49. Reference: NFPA 1041, 4.4.5 and 4.4.5(a)(b)
IFSTA, *Fire and Emergency Services Instructor,* 6th Edition, 4th Printing, page 137.
Answer: B

50. Reference: NFPA 1041, 4.4.5 and 4.4.5(a)(b)
IFSTA, *Fire and Emergency Services Instructor,* 6th Edition, 4th Printing, page 83.
Answer: A

51. Reference: NFPA 1041, 4.4.5 and 4.4.5(a)(b)
IFSTA, *Fire and Emergency Services Instructor,* 6th Edition, 4th Printing, page 83.
Answer: D

52. Reference: NFPA 1041, 4.4.5 and 4.4.5(a)(b)
IFSTA, *Fire and Emergency Services Instructor,* 6th Edition, 4th Printing, page 83.
Answer: C

53. Reference: NFPA 1041, 4.4.5, 4.4.5(a)(b), 4.4.3, and 4.4.3(a)(b)
IFSTA, *Fire and Emergency Services Instructor,* 6th Edition, 4th Printing, page 160.
Answer: B

54. Reference: NFPA 1041, 4.4.5, 4.4.5(a)(b), 4.4.3, and 4.4.3(a)(b)
IFSTA, *Fire and Emergency Services Instructor,* 6th Edition, 4th Printing, page 154.
Answer: D

55. Reference: NFPA 1041, 4.4.6, 4.4.6(a)(b), 4.4.7, and 4.4.7(a)(b)
IFSTA, *Fire and Emergency Services Instructor,* 6th Edition, 4th Printing, pages 231–232.
Answer: D

56. Reference: NFPA 1041, 4.4.6, 4.4.6(a)(b), 4.4.7, and 4.4.7(a)(b)
IFSTA, *Fire and Emergency Services Instructor,* 6th Edition, 4th Printing, page 221.
Answer: D

57. Reference: NFPA 1041, 4.4.6 and 4.4.6(a)(b)
IFSTA, *Fire and Emergency Services Instructor,* 6th Edition, 4th Printing, page 224.
Answer: C

58. Reference: NFPA 1041, 4.4.7 and 4.4.7(a)(b)
IFSTA, *Fire and Emergency Services Instructor,* 6th Edition, 4th Printing, pages 226–227.
Answer: D

59. Reference: NFPA 1041, 4.5.2, 4.5.2(a)(b), 4.5.5, and 4.5.5(a)(b)
IFSTA, *Fire and Emergency Services Instructor,* 6th Edition, 4th Printing, page 94.
Answer: D

60. Reference: NFPA 1041, 4.5.2 and 4.5.2(a)
IFSTA, *Fire and Emergency Services Instructor,* 6th Edition, 4th Printing, page 199.
Answer: C

61. Reference: NFPA 1041, 4.5.2, 4.5.2(a), 4.5.5, and 4.5.5(a)
IFSTA, *Fire and Emergency Services Instructor,* 6th Edition, 4th Printing, page 203.
Answer: A

62. Reference: NFPA 1041, 4.5.2, 4.5.2(a), 4.5.5, and 4.5.5(a)
IFSTA, *Fire and Emergency Services Instructor,* 6th Edition, 4th Printing, page 192.
Answer: A

63. Reference: NFPA 1041, 4.5.2 and 4.5.2(a)
IFSTA, *Fire and Emergency Services Instructor,* 6th Edition, 4th Printing, page 189.
Answer: B

64. Reference: NFPA 1041, 4.5.2, 4.5.2(a), 4.5.3, and 4.5.3(a)
IFSTA, *Fire and Emergency Services Instructor,* 6th Edition, 4th Printing, pages 199–200.
Answer: C

65. Reference: NFPA 1041, 4.5.2 and 4.5.2(a)(b)
IFSTA, *Fire and Emergency Services Instructor,* 6th Edition, 4th Printing, page 94.
Answer: D

66. Reference: NFPA 1041, 4.5.3 and 4.5.3(a)
IFSTA, *Fire and Emergency Services Instructor,* 6th Edition, 4th Printing, page 214.
Answer: B

67. Reference: NFPA 1041, 4.5.3, 4.5.3(a), and 1.3.5
IFSTA, *Fire and Emergency Services Instructor,* 6th Edition, 4th Printing, page 39.
Answer: C

68. Reference: NFPA 1041, 4.5.3, 4.5.3(a), 4.5.4, 4.5.4(a)(b), 4.5.5, and 4.5.5(a)(b)
IFSTA, *Fire and Emergency Services Instructor,* 6th Edition, 4th Printing, page 200.
Answer: C

69. Reference: NFPA 1041, 4.5.4 and 4.5.4(a)
IFSTA, *Fire and Emergency Services Instructor,* 6th Edition, 4th Printing, page 217.
Answer: B

70. Reference: NFPA 1041, 4.5.4 and 4.5.4(a)
IFSTA, *Fire and Emergency Services Instructor,* 6th Edition, 4th Printing, pages 99 and 217.
Answer: C

71. Reference: NFPA 1041, 4.5.4, 4.5.4(a), 4.5.5, and 4.5.5(a)
IFSTA, *Fire and Emergency Services Instructor,* 6th Edition, 4th Printing, page 214.
Answer: A

72. Reference: NFPA 1041, 4.5.5, 4.5.5(a)(b), 4.5.2, and 4.5.2(a)(b)
IFSTA, *Fire and Emergency Services Instructor,* 6th Edition, 4th Printing, page 94.
Answer: A

73. Reference: NFPA 1041, 4.5.5 and 4.5.5(a)
IFSTA, *Fire and Emergency Services Instructor,* 6th Edition, 4th Printing, page 194.
Answer: C

74. Reference: NFPA 1041, 4.5.5 and 4.5.5(a)
IFSTA, *Fire and Emergency Services Instructor,* 6th Edition, 4th Printing, page 217.
Answer: B

75. Reference: NFPA 1041, 4.5.5 and 4.5.5(a)(b)
IFSTA, *Fire and Emergency Services Instructor,* 6th Edition, 4th Printing, pages 215 and 218.
Answer: D

Don't forget to enter the information on your Personal Progress Plotter and answer the Yes and No question at the end of the Examination. This step is extremely important for the successful completion of the Systematic Approach to Examination Preparation!

Examination I-2 Answer Key

Directions

Follow these steps carefully for completing the feedback part of SAEP:

1. After calculating your score, look up the answers for the examination items you missed as well as those on which you guessed, even if you guessed correctly. If you are guessing, it means the answer is not perfectly clear. In this process, we are committed to making you as knowledgeable as possible.
2. Enter the number of missed and guessed examination items in the blanks on your Personal Progress Plotter.
3. Highlight the answer in the reference materials. Read the paragraph preceding and the paragraph following the one in which the correct answer is located. Enter the paragraph number and page number next to the guessed or missed examination item on your examination. Count any part of a paragraph at the beginning of the page as one paragraph until you reach the paragraph containing your highlighted answer. This step will help you locate and review your missed and guessed examination items later in the process. It is essential to learning the material in context and by association. These learning techniques (context/association) are the very backbone of the SAEP approach.
4. Once you have completed the feedback part, you may proceed to the next examination.

1. Reference: NFPA 1041, 4.2.2, 4.2.2(a), 4.3.1, 4.3.3, and 4.3.3(a)(b)
IFSTA, *Fire and Emergency Services Instructor,* 6th Edition, 4th Printing, page 102.
Answer: A

2. Reference: NFPA 1041, 4.2.2 and 4.2.2(a)
IFSTA, *Fire and Emergency Services Instructor,* 6th Edition, 4th Printing, pages 142–143.
Answer: A

3. Reference: NFPA 1041, 4.2.3, 4.2.3(a)(b), 1.3.4, and 1.3.7
IFSTA, *Fire and Emergency Services Instructor,* 6th Edition, 4th Printing, page 179.
Answer: A

4. Reference: NFPA 1041, 4.2.3, 4.2.3(a), 4.5.3, and 4.5.3(a)
IFSTA, *Fire and Emergency Services Instructor,* 6th Edition, 4th Printing, pages 242–243.
Answer: D

5. Reference: NFPA 1041, 4.2.3 and 4.2.3(a)(b)
IFSTA, *Fire and Emergency Services Instructor,* 6th Edition, 4th Printing, pages 28–30.
Answer: B

6. Reference: NFPA 1041, 4.2.3 and 4.2.3(a)(b)
IFSTA, *Fire and Emergency Services Instructor,* 6th Edition, 4th Printing, page 29.
Answer: D

7. Reference: NFPA 1041, 4.3.2 and 4.3.2(a)(b)
IFSTA, *Fire and Emergency Services Instructor,* 6th Edition, 4th Printing, page 157.
Answer: A

8. Reference: NFPA 1041, 4.3.2, 4.3.2(a)(b), 4.3.3, 4.3.3(a), 4.4.3, and 4.4.3(a)(b)
IFSTA, *Fire and Emergency Services Instructor,* 6th Edition, 4th Printing, page 11.
Answer: D

9. Reference: NFPA 1041, 4.3.2, 4.3.2(a)(b), 4.4.2, and 4.4.2(a)
IFSTA, *Fire and Emergency Services Instructor,* 6th Edition, 4th Printing, page 178.
Answer: A

10. Reference: NFPA 1041, 4.3.2, 4.3.2(a)(b), 4.3.3, and 4.3.3(a)(b)
IFSTA, *Fire and Emergency Services Instructor,* 6th Edition, 4th Printing, pages 115–116.
Answer: A

11. Reference: NFPA 1041, 4.3.2, 4.3.2(a)(b), 4.3.3, and 4.3.3(a)(b)
IFSTA, *Fire and Emergency Services Instructor,* 6th Edition, 4th Printing, pages 115 and 118.
Answer: C

12. Reference: NFPA 1041, 4.3.2 and 4.3.2(a)(b)
IFSTA, *Fire and Emergency Services Instructor,* 6th Edition, 4th Printing, pages 169–170.
Answer: A

13. Reference: NFPA 1041, 4.3.2 and 4.3.2(a)(b)
IFSTA, *Fire and Emergency Services Instructor,* 6th Edition, 4th Printing, pages 125–128.
Answer: C

14. Reference: NFPA 1041, 4.3.3, 4.3.3(a)(b), 4.4.3, and 4.4.3(a)(b)
IFSTA, *Fire and Emergency Services Instructor,* 6th Edition, 4th Printing, page 102.
Answer: C

15. Reference: NFPA 1041, 4.3.3 and 4.3.3(a)(b)
IFSTA, *Fire and Emergency Services Instructor,* 6th Edition, 4th Printing, pages 6–7.
Answer: C

16. Reference: NFPA 1041, 4.3.3 and 4.3.3(a)
IFSTA, *Fire and Emergency Services Instructor,* 6th Edition, 4th Printing, page 115.
Answer: D

17. Reference: NFPA 1041, 4.3.3 and 4.3.3(a)(b)
IFSTA, *Fire and Emergency Services Instructor,* 6th Edition, 4th Printing, pages 57–58 and 395.
Answer: C

18. Reference: NFPA 1041, 4.3.3 and 4.3.3(a)(b)
IFSTA, *Fire and Emergency Services Instructor,* 6th Edition, 4th Printing, page 103.
Answer: C

19. Reference: NFPA 1041, 4.3.3 and 4.3.3(a)(b)
IFSTA, *Fire and Emergency Services Instructor,* 6th Edition, 4th Printing, page 113.
Answer: A

20. Reference: NFPA 1041, 4.3.3 and 4.3.3(a)(b)
IFSTA, *Fire and Emergency Services Instructor,* 6th Edition, 4th Printing, page 113.
Answer: D

21. Reference: NFPA 1041, 4.3.3 and 4.3.3(a)(b)
IFSTA, *Fire and Emergency Services Instructor,* 6th Edition, 4th Printing, page 119.
Answer: A

22. Reference: NFPA 1041, 4.3.3, 4.3.3(a)(b), 4.4.3, and 4.4.3(a)(b)
IFSTA, *Fire and Emergency Services Instructor,* 6th Edition, 4th Printing, pages 142–144.
Answer: B

23. Reference: NFPA 1041, 4.3.3 and 4.3.3(a)(b)
IFSTA, *Fire and Emergency Services Instructor,* 6th Edition, 4th Printing, pages 142–144.
Answer: C

24. Reference: NFPA 1041, 4.4.2 and 4.4.2(a)(b)
IFSTA, *Fire and Emergency Services Instructor,* 6th Edition, 4th Printing, page 134.
Answer: D

25. Reference: NFPA 1041, 4.4.2 and 4.4.2(a)(b)
IFSTA, *Fire and Emergency Services Instructor,* 6th Edition, 4th Printing, page 131.
Answer: C

26. Reference: NFPA 1041, 4.4.2 and 4.4.2(a)(b)
IFSTA, *Fire and Emergency Services Instructor,* 6th Edition, 4th Printing, page 24.
Answer: C

27. Reference: NFPA 1041, 4.4.2 and 4.4.2(a)(b)
IFSTA, *Fire and Emergency Services Instructor,* 6th Edition, 4th Printing, page 31.
Answer: D

28. Reference: NFPA 1041, 4.4.2 and 4.4.2(a)
IFSTA, *Fire and Emergency Services Instructor,* 6th Edition, 4th Printing, page 10.
Answer: B

29. Reference: NFPA 1041, 4.4.2 and 4.4.2(a)(b)
IFSTA, *Fire and Emergency Services Instructor,* 6th Edition, 4th Printing, page 166.
Answer: A

30. Reference: NFPA 1041, 4.4.2, 4.4.2(a)(b), 4.4.3, 4.4.3(a)(b), 4.4.4, and 4.4.4(a)
IFSTA, *Fire and Emergency Services Instructor,* 6th Edition, 4th Printing, page 64.
Answer: B

31. Reference: NFPA 1041, 4.4.2 and 4.4.2(a)
IFSTA, *Fire and Emergency Services Instructor,* 6th Edition, 4th Printing, page 151.
Answer: D

32. Reference: NFPA 1041, 4.4.3, 4.4.3(a)(b), 4.5.1, 4.5.3, and 4.5.3(a)
IFSTA, *Fire and Emergency Services Instructor,* 6th Edition, 4th Printing, page 118.
Answer: C

33. Reference: NFPA 1041, 4.4.3 and 4.4.3(a)(b)
IFSTA, *Fire and Emergency Services Instructor,* 6th Edition, 4th Printing, pages 64–65.
Answer: B

34. Reference: NFPA 1041, 4.4.3 and 4.4.3(a)(b)
IFSTA, *Fire and Emergency Services Instructor,* 6th Edition, 4th Printing, page 7.
Answer: D

35. Reference: NFPA 1041, 4.4.3 and 4.4.3(a)(b)
IFSTA, *Fire and Emergency Services Instructor,* 6th Edition, 4th Printing, page 64.
Answer: A

36. Reference: NFPA 1041, 4.4.3, 4.4.3(a)(b), 4.4.5, and 4.4.5(a)(b)
IFSTA, *Fire and Emergency Services Instructor,* 6th Edition, 4th Printing, page 11.
Answer: B

37. Reference: NFPA 1041, 4.4.3 and 4.4.3(a)(b)
IFSTA, *Fire and Emergency Services Instructor,* 6th Edition, 4th Printing, page 74.
Answer: A

38. Reference: NFPA 1041, 4.4.3 and 4.4.3(a)
IFSTA, *Fire and Emergency Services Instructor,* 6th Edition, 4th Printing, page 74.
Answer: D

39. Reference: NFPA 1041, 4.4.3, 4.4.3(a)(b), 4.4.5, and 4.4.5(a)(b)
IFSTA, *Fire and Emergency Services Instructor,* 6th Edition, 4th Printing, page 54.
Answer: C

40. Reference: NFPA 1041, 4.4.3, 4.4.3(a)(b), 4.4.5, and 4.4.5(a)(b)
IFSTA, *Fire and Emergency Services Instructor,* 6th Edition, 4th Printing, pages 64–65.
Answer: C

41. Reference: NFPA 1041, 4.4.3, 4.4.3(a)(b), 4.4.5, and 4.4.5(a)(b)
IFSTA, *Fire and Emergency Services Instructor,* 6th Edition, 4th Printing, pages 53–54.
Answer: B

42. Reference: NFPA 1041, 4.4.3, 4.4.3(a)(b), 4.4.5, and 4.4.5(a)(b)
IFSTA, *Fire and Emergency Services Instructor,* 6th Edition, 4th Printing, page 65.
Answer: B

43. Reference: NFPA 1041, 4.4.3, 4.4.3(a)(b), 4.4.5, and 4.4.5(a)(b)
IFSTA, *Fire and Emergency Services Instructor,* 6th Edition, 4th Printing, page 13.
Answer: B

44. Reference: NFPA 1041, 4.4.3, 4.4.3(a)(b), 4.4.5, and 4.4.5(a)(b)
IFSTA, *Fire and Emergency Services Instructor,* 6th Edition, 4th Printing, page 13.
Answer: D

45. Reference: NFPA 1041, 4.4.4 and 4.4.4(a)
IFSTA, *Fire and Emergency Services Instructor,* 6th Edition, 4th Printing, page 151.
Answer: D

46. Reference: NFPA 1041, 4.4.4 and 4.4.4(a)
IFSTA, *Fire and Emergency Services Instructor,* 6th Edition, 4th Printing, page 131.
Answer: D

47. Reference: NFPA 1041, 4.4.5 and 4.4.5(a)(b)
IFSTA, *Fire and Emergency Services Instructor,* 6th Edition, 4th Printing, page 4.
Answer: B

48. Reference: NFPA 1041, 4.4.5 and 4.4.5(a)(b)
IFSTA, *Fire and Emergency Services Instructor,* 6th Edition, 4th Printing, page 83.
Answer: A

49. Reference: NFPA 1041, 4.4.5 and 4.4.5(a)(b)
IFSTA, *Fire and Emergency Services Instructor,* 6th Edition, 4th Printing, page 137.
Answer: B

50. Reference: NFPA 1041, 4.4.5 and 4.4.5(a)(b)
IFSTA, *Fire and Emergency Services Instructor,* 6th Edition, 4th Printing, page 82.
Answer: D

51. Reference: NFPA 1041, 4.4.5 and 4.4.5(a)(b)
IFSTA, *Fire and Emergency Services Instructor,* 6th Edition, 4th Printing, page 82.
Answer: C

52. Reference: NFPA 1041, 4.4.5 and 4.4.5(a)(b)
IFSTA, *Fire and Emergency Services Instructor,* 6th Edition, 4th Printing, page 83.
Answer: D

53. Reference: NFPA 1041, 4.4.5 and 4.4.5(a)(b)
IFSTA, *Fire and Emergency Services Instructor,* 6th Edition, 4th Printing, page 82
Answer: D

54. Reference: NFPA 1041, 4.4.6, 4.4.6(a)(b), 4.4.7, and 4.4.7(a)(b)
IFSTA, *Fire and Emergency Services Instructor,* 6th Edition, 4th Printing, page 241.
Answer: C

55. Reference: NFPA 1041, 4.4.6, 4.4.6(a)(b), 4.4.7, and 4.4.7(a)(b)
IFSTA, *Fire and Emergency Services Instructor,* 6th Edition, 4th Printing, page 152.
Answer: D

56. Reference: NFPA 1041, 4.4.6, 4.4.6(a)(b), 4.4.7, and 4.4.7(a)(b)
IFSTA, *Fire and Emergency Services Instructor,* 6th Edition, 4th Printing, pages 238–239.
Answer: B

57. Reference: NFPA 1041, 4.4.7 and 4.4.7(a)(b)
IFSTA, *Fire and Emergency Services Instructor,* 6th Edition, 4th Printing, pages 234–236.
Answer: B

58. Reference: NFPA 1041, 4.4.7 and 4.4.7(a)(b)
IFSTA, *Fire and Emergency Services Instructor,* 6th Edition, 4th Printing, pages 143–144 and 152.
Answer: C

59. Reference: NFPA 1041, 4.4.7 and 4.4.7(a)(b)
IFSTA, *Fire and Emergency Services Instructor,* 6th Edition, 4th Printing, page 232.
Answer: C

60. Reference: NFPA 1041, 4.4.7 and 4.4.7(a)(b)
IFSTA, *Fire and Emergency Services Instructor,* 6th Edition, 4th Printing, pages 233 and 236.
Answer: D

61. Reference: NFPA 1041, 4.4.7, 4.4.7(a)(b), 4.4.6, and 4.4.6(a)(b)
IFSTA, *Fire and Emergency Services Instructor,* 6th Edition, 4th Printing, page 152.
Answer: D

62. Reference: NFPA 1041, 4.4.7 and 4.4.7(a)(b)
IFSTA, *Fire and Emergency Services Instructor,* 6th Edition, 4th Printing, page 224.
Answer: C

63. Reference: NFPA 1041, 4.5.2, 4.5.2(a)(b), 4.5.4, 4.5.4(a)(b), 4.5.5, and 4.5.5(a)(b)
IFSTA, *Fire and Emergency Services Instructor,* 6th Edition, 4th Printing, page 178.
Answer: D

64. Reference: NFPA 1041, 4.5.2, 4.5.2(a), 4.5.3, and 4.5.3(a)
IFSTA, *Fire and Emergency Services Instructor,* 6th Edition, 4th Printing, pages 199–200.
Answer: D

65. Reference: NFPA 1041, 4.5.2, 4.5.2(a), 4.5.3, and 4.5.3(a)
IFSTA, *Fire and Emergency Services Instructor,* 6th Edition, 4th Printing, page 199.
Answer: C

66. Reference: NFPA 1041, 4.5.2, 4.5.2(a)(b), 4.5.3, 4.5.3(a), 4.5.4, 4.5.4(a)(b), 4.5.5, and 4.5.5(a)(b)
IFSTA, *Fire and Emergency Services Instructor,* 6th Edition, 4th Printing, page 217.
Answer: A

67. Reference: NFPA 1041, 4.5.3 and 4.5.3(a)
IFSTA, *Fire and Emergency Services Instructor,* 6th Edition, 4th Printing, page 208.
Answer: A

68. Reference: NFPA 1041, 4.5.3, 4.5.3(a), 4.5.2, and 4.5.2(a)
IFSTA, *Fire and Emergency Services Instructor,* 6th Edition, 4th Printing, pages 193–194.
Answer: C

69. Reference: NFPA 1041, 4.5.3, 4.5.3(a), and 1.3.5
IFSTA, *Fire and Emergency Services Instructor,* 6th Edition, 4th Printing, page 39.
Answer: C

70. Reference: NFPA 1041, 4.5.4 and 4.5.4(a)(b)
IFSTA, *Fire and Emergency Services Instructor,* 6th Edition, 4th Printing, page 136.
Answer: C

71. Reference: NFPA 1041, 4.5.4 and 4.5.4(a)(b)
IFSTA, *Fire and Emergency Services Instructor,* 6th Edition, 4th Printing, page 136.
Answer: C

72. Reference: NFPA 1041, 4.5.4, 4.5.4(a), and 1.3.5
IFSTA, *Fire and Emergency Services Instructor,* 6th Edition, 4th Printing, page 39.
Answer: B

73. Reference: NFPA 1041, 4.5.4 and 4.5.4(a)
IFSTA, *Fire and Emergency Services Instructor,* 6th Edition, 4th Printing, page 215.
Answer: B

74. Reference: NFPA 1041, 4.5.5 and 4.5.5(a)(b)
IFSTA, *Fire and Emergency Services Instructor,* 6th Edition, 4th Printing, page 118.
Answer: B

75. Reference: NFPA 1041, 4.5.5 and 4.5.5(a)(b)
IFSTA, *Fire and Emergency Services Instructor,* 6th Edition, 4th Printing, page 82.
Answer: B

Don't forget to enter the information on your Personal Progress Plotter and answer the Yes and No question at the end of the Examination. This step is extremely important for the successful completion of the Systematic Approach to Examination Preparation!

Examination I-3 Answer Key

Directions

Follow these steps carefully for completing the feedback part of SAEP:

1. After calculating your score, look up the answers for the examination items you missed as well as those on which you guessed, even if you guessed correctly. If you are guessing, it means the answer is not perfectly clear. In this process, we are committed to making you as knowledgeable as possible.
2. Enter the number of missed and guessed examination items in the blanks on your Personal Progress Plotter.
3. Highlight the answer in the reference materials. Read the paragraph preceding and the paragraph following the one in which the correct answer is located. Enter the paragraph number and page number next to the guessed or missed examination item on your examination. Count any part of a paragraph at the beginning of the page as one paragraph until you reach the paragraph containing your highlighted answer. This step will help you locate and review your missed and guessed examination items later in the process. It is essential to learning the material in context and by association. These learning techniques (context/association) are the very backbone of the SAEP approach.
4. Congratulations! You have completed the examination and feedback steps of SAEP when you have highlighted your guessed and missed examination items for this examination.

Proceed to Phases II and III. Study the materials carefully in these important phases-they will help you polish your examination-taking skills. Approximately two to three days before you take your next examination, carefully read all of the highlighted information in the reference materials using the same techniques you applied during the feedback step. This will reinforce your learning and provide you with an added level of confidence going into the examination.

Someone once said to professional golfer Tom Watson after he won several tournament championships, "You are really lucky to have won those championships. You are really on a streak." Watson was reported to have replied, "Yes, there is some luck involved, but what I have really noticed is that the more I practice, the luckier I get." What Watson was saying is that good luck usually results from good preparation. This line of thinking certainly applies to learning the rules and hints of examination taking.

Rule 7

Good luck = good preparation.

1. Reference: NFPA 1041, 4.1.1
   IFSTA, *Fire and Emergency Services Instructor,* 6th Edition, 4th Printing, page vii.
   Answer: B

2. Reference: NFPA 1041, 4.1.1, 4.3.1, and 4.4.1
   IFSTA, *Fire and Emergency Services Instructor,* 6th Edition, 4th Printing, pages 3–4.
   Answer: A

3. Reference: NFPA 1041, 4.2.2, 4.2.2(a), 4.3.1, 4.3.3, and 4.3.3(a)(b)
IFSTA, *Fire and Emergency Services Instructor,* 6th Edition, 4th Printing, page 12.
Answer: A

4. Reference: NFPA 1041, 4.2.2 and 4.2.2(a)
IFSTA, *Fire and Emergency Services Instructor,* 6th Edition, 4th Printing, pages 142–143.
Answer: A

5. Reference: NFPA 1041, 4.2.3, 4.2.3(a)(b), 1.3.4, and 1.3.7
IFSTA, *Fire and Emergency Services Instructor,* 6th Edition, 4th Printing, page 179.
Answer: A

6. Reference: NFPA 1041, 4.2.3 and 4.2.3(a)(b)
IFSTA, *Fire and Emergency Services Instructor,* 6th Edition, 4th Printing, page 252.
Answer: C

7. Reference: NFPA 1041, 4.2.3, 4.2.3(a)(b), 1.3.5, and 1.3.7
IFSTA, *Fire and Emergency Services Instructor,* 6th Edition, 4th Printing, page 181.
Answer: A

8. Reference: NFPA 1041, 4.2.3, 4.2.3(a), 4.5.3, and 4.5.3(a)
IFSTA, *Fire and Emergency Services Instructor,* 6th Edition, 4th Printing, pages 242–243.
Answer: D

9. Reference: NFPA 1041, 4.2.3 and 4.2.3(a)(b)
IFSTA, *Fire and Emergency Services Instructor,* 6th Edition, 4th Printing, page 29.
Answer: B

10. Reference: NFPA 1041, 4.2.3 and 4.2.3(a)(b)
IFSTA, *Fire and Emergency Services Instructor,* 6th Edition, 4th Printing, pages 28–30.
Answer: B

11. Reference: NFPA 1041, 4.3.2, 4.3.2(a)(b), and 4.3.1
IFSTA, *Fire and Emergency Services Instructor,* 6th Edition, 4th Printing, page 98.
Answer: A

12. Reference: NFPA 1041, 4.3.2 and 4.3.2(a)
IFSTA, *Fire and Emergency Services Instructor,* 6th Edition, 4th Printing, pages 68–69.
Answer: D

13. Reference: NFPA 1041, 4.3.2 and 4.3.2(a)(b)
IFSTA, *Fire and Emergency Services Instructor,* 6th Edition, 4th Printing, page 222.
Answer: B

14. Reference: NFPA 1041, 4.3.2 and 4.3.2(a)(b)
IFSTA, *Fire and Emergency Services Instructor,* 6th Edition, 4th Printing, page 96.
Answer: C

15. Reference: NFPA 1041, 4.3.2, 4.3.2(a)(b), 4.3.3, and 4.3.3(a)(b)
IFSTA, *Fire and Emergency Services Instructor,* 6th Edition, 4th Printing, pages 115 and 118.
Answer: C

16. Reference: NFPA 1041, 4.3.2, 4.3.2(a), 4.4.2, 4.4.2(a)(b), 4.4.3, and 4.4.3(a)(b)
IFSTA, *Fire and Emergency Services Instructor,* 6th Edition, 4th Printing, pages 117 and 156.
Answer: A

17. Reference: NFPA 1041, 4.3.3 and 4.3.3(a)(b)
IFSTA, *Fire and Emergency Services Instructor,* 6th Edition, 4th Printing, pages 117–118.
Answer: C

18. Reference: NFPA 1041, 4.3.3 and 4.3.3(a)(b)
IFSTA, *Fire and Emergency Services Instructor,* 6th Edition, 4th Printing, pages 6–7.
Answer: C

19. Reference: NFPA 1041, 4.3.3, 4.3.3(a)(b), and 3.3.2.1
IFSTA, *Fire and Emergency Services Instructor,* 6th Edition, 4th Printing, pages 6–8.
Answer: D

20. Reference: NFPA 1041, 4.3.3 and 4.3.3(a)(b)
IFSTA, *Fire and Emergency Services Instructor,* 6th Edition, 4th Printing, pages 96 and 390.
Answer: C

21. Reference: NFPA 1041, 4.3.3 and 4.3.3(a)(b)
IFSTA, *Fire and Emergency Services Instructor,* 6th Edition, 4th Printing, pages 57–58 and 395.
Answer: C

22. Reference: NFPA 1041, 4.3.3 and 4.3.3(a)(b)
IFSTA, *Fire and Emergency Services Instructor,* 6th Edition, 4th Printing, page 113.
Answer: A

23. Reference: NFPA 1041, 4.3.3 and 4.3.3(a)(b)
IFSTA, *Fire and Emergency Services Instructor,* 6th Edition, 4th Printing, page 6.
Answer: A

24. Reference: NFPA 1041, 4.3.3 and 4.3.3(a)(b)
IFSTA, *Fire and Emergency Services Instructor,* 6th Edition, 4th Printing, page 119.
Answer: A

25. Reference: NFPA 1041, 4.3.3 and 4.3.3(a)(b)
IFSTA, *Fire and Emergency Services Instructor,* 6th Edition, 4th Printing, page 120.
Answer: D

26. Reference: NFPA 1041, 4.3.3 and 4.3.3(a)(b)
IFSTA, *Fire and Emergency Services Instructor,* 6th Edition, 4th Printing, page 96.
Answer: A

27. Reference: NFPA 1041, 4.3.3 and 4.3.3(a)(b)
IFSTA, *Fire and Emergency Services Instructor,* 6th Edition, 4th Printing, page 74.
Answer: A

28. Reference: NFPA 1041, 4.3.3 and 4.3.3(a)(b)
IFSTA, *Fire and Emergency Services Instructor,* 6th Edition, 4th Printing, pages 104 and 390.
Answer: C

29. Reference: NFPA 1041, 4.3.3, 4.3.3(a)(b), 4.4.3, and 4.4.3(a)(b)
IFSTA, *Fire and Emergency Services Instructor,* 6th Edition, 4th Printing, page 7.
Answer: A

30. Reference: NFPA 1041, 4.3.3 and 4.3.3(a)(b)
IFSTA, *Fire and Emergency Services Instructor,* 6th Edition, 4th Printing, page 96.
Answer: A

31. Reference: NFPA 1041, 4.4.2, 4.4.2(a)(b), 4.4.3, and 4.4.3(a)(b)
IFSTA, *Fire and Emergency Services Instructor,* 6th Edition, 4th Printing, page 93.
Answer: C

32. Reference: NFPA 1041, 4.4.2 and 4.4.2(a)
IFSTA, *Fire and Emergency Services Instructor,* 6th Edition, 4th Printing, page 178.
Answer: C

33. Reference: NFPA 1041, 4.4.2, 4.4.2(a)(b), 4.4.5, and 4.4.5(a)(b)
IFSTA, *Fire and Emergency Services Instructor,* 6th Edition, 4th Printing, page 29.
Answer: A

34. Reference: NFPA 1041, 4.4.2, 4.4.2(a)(b) 4.4.3, and 4.4.3(a)(b)
IFSTA, *Fire and Emergency Services Instructor,* 6th Edition, 4th Printing, page 118.
Answer: C

35. Reference: NFPA 1041, 4.4.2 and 4.4.2(a)(b)
IFSTA, *Fire and Emergency Services Instructor,* 6th Edition, 4th Printing, page 149.
Answer: C

36. Reference: NFPA 1041, 4.4.2 and 4.4.2(a)(b)
IFSTA, *Fire and Emergency Services Instructor,* 6th Edition, 4th Printing, page 24.
Answer: C

37. Reference: NFPA 1041, 4.4.2 and 4.4.2(a)(b)
IFSTA, *Fire and Emergency Services Instructor,* 6th Edition, 4th Printing, pages 24 and 146.
Answer: C

38. Reference: NFPA 1041, 4.4.2 and 4.4.2(a)(b)
IFSTA, *Fire and Emergency Services Instructor,* 6th Edition, 4th Printing, page 31.
Answer: D

39. Reference: NFPA 1041, 4.4.2 and 4.4.2(a)(b)
IFSTA, *Fire and Emergency Services Instructor,* 6th Edition, 4th Printing, page 148.
Answer: D

40. Reference: NFPA 1041, 4.4.2, 4.4.2(a)(b), 4.4.4, and 4.4.4(a)
IFSTA, *Fire and Emergency Services Instructor,* 6th Edition, 4th Printing, pages 145–148.
Answer: D

41. Reference: NFPA 1041, 4.4.2 and 4.4.2(a)(b)
IFSTA, *Fire and Emergency Services Instructor,* 6th Edition, 4th Printing, page 166.
Answer: A

42. Reference: NFPA 1041, 4.4.2, 4.4.2(a)(b), 4.4.3, 4.4.3(a)(b), 4.4.4, and 4.4.4(a)
IFSTA, *Fire and Emergency Services Instructor,* 6th Edition, 4th Printing, page 64.
Answer: B

43. Reference: NFPA 1041, 4.4.2 and 4.4.2(a)(b)
IFSTA, *Fire and Emergency Services Instructor,* 6th Edition, 4th Printing, pages 164 and 396.
Answer: B

44. Reference: NFPA 1041, 4.4.2, 4.4.2(a)(b), 4.4.3, 4.4.3(a)(b), 4.4.4, 4.4.4(a), 4.4.5, 4.4.5(a)(b), 4.4.6, 4.4.6(a)(b), 4.4.7, and 4.4.7(a)(b)
IFSTA, *Fire and Emergency Services Instructor,* 6th Edition, 4th Printing, pages 230 and 238.
Answer: A

45. Reference: NFPA 1041, 4.4.2 and 4.4.2(a)
IFSTA, *Fire and Emergency Services Instructor,* 6th Edition, 4th Printing, page 151.
Answer: D

46. Reference: NFPA 1041, 4.4.3, 4.4.3(a)(b), 4.5.1, 4.5.3, and 4.5.3(a)
IFSTA, *Fire and Emergency Services Instructor,* 6th Edition, 4th Printing, page 118.
Answer: C

47. Reference: NFPA 1041, 4.4.3 and 4.4.3(a)(b)
IFSTA, *Fire and Emergency Services Instructor,* 6th Edition, 4th Printing, page 158.
Answer: C

48. Reference: NFPA 1041, 4.4.3 and 4.4.3(a)(b)
IFSTA, *Fire and Emergency Services Instructor,* 6th Edition, 4th Printing, pages 64–65.
Answer: B

49. Reference: NFPA 1041, 4.4.3 and 4.4.3(a)(b)
IFSTA, *Fire and Emergency Services Instructor,* 6th Edition, 4th Printing, page 64.
Answer: A

50. Reference: NFPA 1041, 4.4.3 and 4.4.3(a)(b)
IFSTA, *Fire and Emergency Services Instructor,* 6th Edition, 4th Printing, pages 68–69 and 153.
Answer: B

51. Reference: NFPA 1041, 4.4.3 and 4.4.3(a)(b)
IFSTA, *Fire and Emergency Services Instructor,* 6th Edition, 4th Printing, page 106.
Answer: B

52. Reference: NFPA 1041, 4.4.3, 4.4.3(a)(b), 4.4.4, and 4.4.4(a)
IFSTA, *Fire and Emergency Services Instructor,* 6th Edition, 4th Printing, page 166.
Answer: D

53. Reference: NFPA 1041, 4.4.3 and 4.4.3(a)(b)
IFSTA, *Fire and Emergency Services Instructor,* 6th Edition, 4th Printing, page 74.
Answer: A

54. Reference: NFPA 1041, 4.4.3 and 4.4.3(a)
IFSTA, *Fire and Emergency Services Instructor,* 6th Edition, 4th Printing, page 74.
Answer: B

55. Reference: NFPA 1041, 4.4.3, 4.4.3(a)(b), 4.4.5, and 4.4.5(a)(b)
IFSTA, *Fire and Emergency Services Instructor,* 6th Edition, 4th Printing, pages 64–65.
Answer: C

56. Reference: NFPA 1041, 4.4.3, 4.4.3(a)(b), 4.4.5, and 4.4.5(a)(b)
IFSTA, *Fire and Emergency Services Instructor,* 6th Edition, 4th Printing, page 53.
Answer: B

57. Reference: NFPA 1041, 4.4.3, 4.4.3(a)(b), 4.4.5, and 4.4.5(a)(b)
IFSTA, *Fire and Emergency Services Instructor,* 6th Edition, 4th Printing, page 65.
Answer: C

58. Reference: NFPA 1041, 4.4.3, 4.4.3(a)(b), 4.4.5, and 4.4.5(a)(b)
IFSTA, *Fire and Emergency Services Instructor,* 6th Edition, 4th Printing, page 221.
Answer: C

59. Reference: NFPA 1041, 4.4.3, 4.4.3(a)(b), 4.4.5, and 4.4.5(a)(b)
IFSTA, *Fire and Emergency Services Instructor,* 6th Edition, 4th Printing, page 10.
Answer: A

60. Reference: NFPA 1041, 4.4.3 and 4.4.3(a)(b)
IFSTA, *Fire and Emergency Services Instructor,* 6th Edition, 4th Printing, page 8.
Answer: B

61. Reference: NFPA 1041, 4.4.4 and 4.4.4(a)
IFSTA, *Fire and Emergency Services Instructor,* 6th Edition, 4th Printing, page 151.
Answer: D

62. Reference: NFPA 1041, 4.4.4 and 4.4.4(a)
IFSTA, *Fire and Emergency Services Instructor,* 6th Edition, 4th Printing, page 131.
Answer: D

63. Reference: NFPA 1041, 4.4.4 and 4.4.4(a)
IFSTA, *Fire and Emergency Services Instructor,* 6th Edition, 4th Printing, page 9.
Answer: D

64. Reference: NFPA 1041, 4.4.4 and 4.4.4(a)
IFSTA, *Fire and Emergency Services Instructor,* 6th Edition, 4th Printing, page 144.
Answer: A

65. Reference: NFPA 1041, 4.4.4 and 4.4.4(a)
IFSTA, *Fire and Emergency Services Instructor,* 6th Edition, 4th Printing, page 144.
Answer: C

66. Reference: NFPA 1041, 4.4.5 and 4.4.5(a)(b)
IFSTA, *Fire and Emergency Services Instructor,* 6th Edition, 4th Printing, page 136.
Answer: D

67. Reference: NFPA 1041, 4.4.5 and 4.4.5(a)(b)
IFSTA, *Fire and Emergency Services Instructor,* 6th Edition, 4th Printing, page 137.
Answer: B

68. Reference: NFPA 1041, 4.4.5 and 4.4.5(a)(b)
IFSTA, *Fire and Emergency Services Instructor,* 6th Edition, 4th Printing, page 83.
Answer: A

69. Reference: NFPA 1041, 4.4.5 and 4.4.5(a)(b)
IFSTA, *Fire and Emergency Services Instructor,* 6th Edition, 4th Printing, page 137.
Answer: B

70. Reference: NFPA 1041, 4.4.5 and 4.4.5(a)(b)
IFSTA, *Fire and Emergency Services Instructor,* 6th Edition, 4th Printing, page 5.
Answer: D

71. Reference: NFPA 1041, 4.4.5 and 4.4.5(a)(b)
IFSTA, *Fire and Emergency Services Instructor,* 6th Edition, 4th Printing, page 82.
Answer: D

72. Reference: NFPA 1041, 4.4.5 and 4.4.5(a)(b)
IFSTA, *Fire and Emergency Services Instructor,* 6th Edition, 4th Printing, page 83.
Answer: D

73. Reference: NFPA 1041, 4.4.5 and 4.4.5(a)(b)
IFSTA, *Fire and Emergency Services Instructor,* 6th Edition, 4th Printing, page 83.
Answer: D

74. Reference: NFPA 1041, 4.4.5 and 4.4.5(a)(b)
IFSTA, *Fire and Emergency Services Instructor,* 6th Edition, 4th Printing, page 83.
Answer: C

75. Reference: NFPA 1041, 4.4.5, 4.4.5(a)(b), 4.4.3, and 4.4.3(a)(b)
IFSTA, *Fire and Emergency Services Instructor,* 6th Edition, 4th Printing, page 160.
Answer: B

76. Reference: NFPA 1041, 4.4.5, 4.4.5(a)(b), 4.4.3, and 4.4.3(a)(b)
IFSTA, *Fire and Emergency Services Instructor,* 6th Edition, 4th Printing, page 154.
Answer: D

77. Reference: NFPA 1041, 4.4.5 and 4.4.5(a)(b)
IFSTA, *Fire and Emergency Services Instructor,* 6th Edition, 4th Printing, page 82
Answer: D

78. Reference: NFPA 1041, 4.4.6, 4.4.6(a)(b), 4.4.7, and 4.4.7(a)(b)
IFSTA, *Fire and Emergency Services Instructor,* 6th Edition, 4th Printing, page 241.
Answer: C

79. Reference: NFPA 1041, 4.4.6, 4.4.6(a)(b), 4.4.7, and 4.4.7(a)(b)
IFSTA, *Fire and Emergency Services Instructor,* 6th Edition, 4th Printing, page 152.
Answer: D

80. Reference: NFPA 1041, 4.4.6, 4.4.6(a)(b), 4.4.7, and 4.4.7(a)(b)
IFSTA, *Fire and Emergency Services Instructor,* 6th Edition, 4th Printing, pages 231–232.
Answer: D

81. Reference: NFPA 1041, 4.4.6, 4.4.6(a)(b), 4.4.7, and 4.4.7(a)(b)
IFSTA, *Fire and Emergency Services Instructor,* 6th Edition, 4th Printing, pages 238–239.
Answer: B

82. Reference: NFPA 1041, 4.4.6 and 4.4.6(a)(b)
IFSTA, *Fire and Emergency Services Instructor,* 6th Edition, 4th Printing, page 224.
Answer: C

83. Reference: NFPA 1041, 4.4.7 and 4.4.7(a)(b)
IFSTA, *Fire and Emergency Services Instructor,* 6th Edition, 4th Printing, pages 143-144 and 152.
Answer: C

84. Reference: NFPA 1041, 4.4.7 and 4.4.7(a)(b)
IFSTA, *Fire and Emergency Services Instructor,* 6th Edition, 4th Printing, page 226.
Answer: B

85. Reference: NFPA 1041, 4.4.7 and 4.4.7(a)(b)
IFSTA, *Fire and Emergency Services Instructor,* 6th Edition, 4th Printing, pages 226–227.
Answer: D

86. Reference: NFPA 1041, 4.5.2, 4.5.2(a)(b), 4.5.5, and 4.5.5(a)(b)
IFSTA, *Fire and Emergency Services Instructor,* 6th Edition, 4th Printing, page 94.
Answer: D

87. Reference: NFPA 1041, 4.5.2, 4.5.2(a), 4.5.3, and 4.5.3(a)
IFSTA, *Fire and Emergency Services Instructor,* 6th Edition, 4th Printing, page 199.
Answer: C

88. Reference: NFPA 1041, 4.5.2 and 4.5.2(a)
IFSTA, *Fire and Emergency Services Instructor,* 6th Edition, 4th Printing, page 199.
Answer: C

89. Reference: NFPA 1041, 4.5.2, 4.5.2(a), 4.5.5, and 4.5.5(a)
IFSTA, *Fire and Emergency Services Instructor,* 6th Edition, 4th Printing, page 192.
Answer: A

90. Reference: NFPA 1041, 4.5.2, 4.5.2(a)(b), 4.5.3, 4.5.3(a), 4.5.4, 4.5.4(a)(b), 4.5.5, and 4.5.5(a)(b)
IFSTA, *Fire and Emergency Services Instructor,* 6th Edition, 4th Printing, page 215.
Answer: B

91. Reference: NFPA 1041, 4.5.2 and 4.5.2(a)
IFSTA, *Fire and Emergency Services Instructor,* 6th Edition, 4th Printing, page 189.
Answer: B

92. Reference: NFPA 1041, 4.5.2, 4.5.2(a), 4.5.3, and 4.5.3(a)
IFSTA, *Fire and Emergency Services Instructor,* 6th Edition, 4th Printing, pages 199–200.
Answer: C

93. Reference: NFPA 1041, 4.5.4 and 4.5.4(a)(b)
IFSTA, *Fire and Emergency Services Instructor,* 6th Edition, 4th Printing, page 136.
Answer: C

94. Reference: NFPA 1041, 4.5.4, 4.5.4(a), and 1.3.5
IFSTA, *Fire and Emergency Services Instructor,* 6th Edition, 4th Printing, page 39.
Answer: B

95. Reference: NFPA 1041, 4.5.4, 4.5.4(a), and 1.3.5
IFSTA, *Fire and Emergency Services Instructor,* 6th Edition, 4th Printing, pages 39 and 253.
Answer: C

96. Reference: NFPA 1041, 4.5.4 and 4.5.4(a)
IFSTA, *Fire and Emergency Services Instructor,* 6th Edition, 4th Printing, pages 99 and 217.
Answer: C

97. Reference: NFPA 1041, 4.5.4 and 4.5.4(a)
IFSTA, *Fire and Emergency Services Instructor,* 6th Edition, 4th Printing, page 215.
Answer: B

98. Reference: NFPA 1041, 4.5.5 and 4.5.5(a)(b)
IFSTA, *Fire and Emergency Services Instructor,* 6th Edition, 4th Printing, page 82.
Answer: B

99. Reference: NFPA 1041, 4.5.5 and 4.5.5(a)
IFSTA, *Fire and Emergency Services Instructor,* 6th Edition, 4th Printing, page 217.
Answer: B

100. Reference: NFPA 1041, 4.5.5 and 4.5.5(a)(b)
IFSTA, *Fire and Emergency Services Instructor,* 6th Edition, 4th Printing, pages 215 and 218.
Answer: D

Don't forget to enter the information on your Personal Progress Plotter and answer the Yes and No question at the end of the Examination. This step is extremely important for the successful completion of the Systematic Approach to Examination Preparation!

APPENDIX B

Examination II-1 Answer Key

Directions

Follow these steps carefully for completing the feedback part of SAEP:

1. After calculating your score, look up the answers for the examination items you missed as well as those on which you guessed, even if you guessed correctly. If you are guessing, it means the answer is not perfectly clear. In this process, we are committed to making you as knowledgeable as possible.
2. Enter the number of missed and guessed examination items in the blanks on your Personal Progress Plotter.
3. Highlight the answer in the reference materials. Read the paragraph preceding and the paragraph following the one in which the correct answer is located. Enter the paragraph number and page number next to the guessed or missed examination item on your examination. Count any part of a paragraph at the beginning of the page as one paragraph until you reach the paragraph containing your highlighted answer. This step will help you locate and review your missed and guessed examination items later in the process. It is essential to learning the material in context and by association. These learning techniques (context/association) are the very backbone of the SAEP approach.
4. Once you have completed the feedback step, you may proceed to the next examination.

1. Reference: NFPA 1041, 5.2.2(a)
   IFSTA, *Fire and Emergency Services Instructor,* 6th Edition, 4th Printing, page 259.
   Answer: C

2. Reference: NFPA 1041, 5.2.2(a)
   IFSTA, *Fire and Emergency Services Instructor,* 6th Edition, 4th Printing, page 107.
   Answer: A

3. Reference: NFPA 1041, 5-2.2(a), 5.2.1 and 5.2.5(a)
   IFSTA, *Fire and Emergency Services Instructor,* 6th Edition, 4th Printing, page 259.
   Answer: A

4. Reference: NFPA 1041, 5.2.2(a) and 5.2.1
   IFSTA, *Fire and Emergency Services Instructor,* 6th Edition, 4th Printing, pages 18–19.
   Answer: D

5. Reference: NFPA 1041, 5.2.3(a)(b)
IFSTA, *Fire and Emergency Services Instructor,* 6th Edition, 4th Printing, page 101.
Answer: D

6. Reference: NFPA 1041, 5.2.3(a)
IFSTA, *Fire and Emergency Services Instructor,* 6th Edition, 4th Printing, page 246.
Answer: A

7. Reference: NFPA 1041, 5.2.3(a)(b) and 5.2.4(a)(b)
IFSTA, *Fire and Emergency Services Instructor,* 6th Edition, 4th Printing, page 262.
Answer: A

8. Reference: NFPA 1041, 5.2.3(a) and 5.2.5(a)(b)
IFSTA, *Company Officer*, Third Edition, page 262.
Answer: C

9. Reference: NFPA 1041, 5.2.4(a)
IFSTA, *Fire and Emergency Services Instructor,* 6th Edition, 4th Printing, page 263.
Answer: D

10. Reference: NFPA 1041, 5.2.4(a) and 5.2.3(a)(b)
IFSTA, *Fire and Emergency Services Instructor,* 6th Edition, 4th Printing, page 262.
Answer: B

11. Reference: NFPA 1041, 5.2.4(a)(b) and 5.2.3(a)(b)
IFSTA, *Fire and Emergency Services Instructor,* 6th Edition, 4th Printing, page 262.
Answer: A

12. Reference: NFPA 1041, 5.2.5(a) and 5.2.2(a)
IFSTA, *Fire and Emergency Services Instructor,* 6th Edition, 4th Printing, page 44.
Answer: D

13. Reference: NFPA 1041, 5.2.5(a), 5.2.6(a), and 5.4.3(a)
IFSTA, *Fire and Emergency Services Instructor,* 6th Edition, 4th Printing, page 248.
Answer: B

14. Reference: NFPA 1041, 5.2.5(a)
IFSTA, *Fire and Emergency Services Instructor,* 6th Edition, 4th Printing, page 133.
Answer: B

15. Reference: NFPA 1041, 5.2.5(a)
IFSTA, *Fire and Emergency Services Instructor,* 6th Edition, 4th Printing, page 242.
Answer: D

16. Reference: NFPA 1041, 5.2.5(a)
IFSTA, *Fire and Emergency Services Instructor,* 6th Edition, 4th Printing, page 44.
Answer: C

17. Reference: NFPA 1041, 5.2.6(a)(b)
IFSTA, *Fire and Emergency Services Instructor,* 6th Edition, 4th Printing, pages 257–258.
Answer: B

18. Reference: NFPA 1041, 5.2.6(a)(b)
IFSTA, *Fire and Emergency Services Instructor,* 6th Edition, 4th Printing, page 49.
Answer: A

19. Reference: NFPA 1041, 5.2.6(a)(b), 5.4.3(a)(b), and 5.2.2(a)
IFSTA, *Fire and Emergency Services Instructor,* 6th Edition, 4th Printing, page 65.
Answer: D

20. Reference: NFPA 1041, 5.2.6(a)(b), 5.2.3(a)(b), and 5.2.4(a)(b)
IFSTA, *Fire and Emergency Services Instructor,* 6th Edition, 4th Printing, page 260.
Answer: C

21. Reference: NFPA 1041, 5.3.2(a)(b) and 5.3.3(a)(b)
IFSTA, *Fire and Emergency Services Instructor,* 6th Edition, 4th Printing, page 95.
Answer: D

22. Reference: NFPA 1041, 5.3.2(a)(b) and 5.3.3(a)(b)
IFSTA, *Fire and Emergency Services Instructor,* 6th Edition, 4th Printing, pages 102–103.
Answer: C

23. Reference: NFPA 1041, 5.3.2(a)(b)
IFSTA, *Fire and Emergency Services Instructor,* 6th Edition, 4th Printing, pages 119–122.
Answer: A

24. Reference: NFPA 1041, 5.3.2(a)(b) and 5.3.3(a)(b)
IFSTA, *Fire and Emergency Services Instructor,* 6th Edition, 4th Printing, page 96.
Answer: C

25. Reference: NFPA 1041, 5.3.2, 5.3.2(a)(b), and 5.3.3(a)(b)
IFSTA, *Fire and Emergency Services Instructor,* 6th Edition, 4th Printing, page 103.
Answer: D

26. Reference: NFPA 1041, 5.3.2, 5.3.2(a)(b), and 5.3.3(a)(b)
IFSTA, *Fire and Emergency Services Instructor,* 6th Edition, 4th Printing, pages 103–104.
Answer: B

27. Reference: NFPA 1041, 5.3.2, 5.3.2(a)(b), and 5.3.3(a)(b)
IFSTA, *Fire and Emergency Services Instructor,* 6th Edition, 4th Printing, pages 102 and 194.
Answer: D

28. Reference: NFPA 1041, 5.3.2 and 5.3.2(a)(b)
IFSTA, *Fire and Emergency Services Instructor,* 6th Edition, 4th Printing, page 104.
Answer: A

29. Reference: NFPA 1041, 5.3.2, 5.3.2.(a)(b), and 5.3.3(a)(b)
IFSTA, *Fire and Emergency Services Instructor,* 6th Edition, 4th Printing, pages 238–239.
Answer: A

30. Reference: NFPA 1041, 5.3.2, 5.3.2(a)(b), and 5.3.3(a)(b)
IFSTA, *Fire and Emergency Services Instructor,* 6th Edition, 4th Printing, pages 236–237.
Answer: D

31. Reference: NFPA 1041, 5.3.2, 5.3.2(a)(b), 5.3.3, and 5.3.3(a)(b)
IFSTA, *Fire and Emergency Services Instructor,* 6th Edition, 4th Printing, page 226.
Answer: B

32. Reference: NFPA 1041, 5.3.2(a)(b) and 5.3.3(a)(b)
IFSTA, *Fire and Emergency Services Instructor,* 6th Edition, 4th Printing, page 104.
Answer: A

33. Reference: NFPA 1041, 5.3.2(a)(b) and 5.3.3(a)(b)
IFSTA, *Fire and Emergency Services Instructor,* 6th Edition, 4th Printing, page 61.
Answer: C

34. Reference: NFPA 1041, 5.3.2(a)(b) and 5.3.3(a)(b)
IFSTA, *Fire and Emergency Services Instructor,* 6th Edition, 4th Printing, pages 233–235.
Answer: C

35. Reference: NFPA 1041, 5.3.2 and 5.3.2(a)(b)
IFSTA, *Fire and Emergency Services Instructor* 6th Edition, 2nd Printing, pages 95-97.
Answer: B

36. Reference: NFPA 1041, 5.3.2, 5.3.2(a)(b), 5.3.3, and 5.3.3(a)(b)
IFSTA, *Fire and Emergency Services Instructor,* 6th Edition, 4th Printing, pages 165–166.
Answer: B

37. Reference: NFPA 1041, 5.3.2 and 5.3.2(a)(b)
IFSTA, *Fire and Emergency Services Instructor,* 6th Edition, 4th Printing, page 55.
Answer: A

38. Reference: NFPA 1041, 5.3.2(a)(b) and 5.3.3(a)(b)
IFSTA, *Fire and Emergency Services Instructor,* 6th Edition, 4th Printing, page 119.
Answer: D

39. Reference: NFPA 1041, 5.3.2(a) and 5.3.3(a)
IFSTA, *Fire and Emergency Services Instructor,* 6th Edition, 4th Printing, pages 116–118.
Answer: D

40. Reference: NFPA 1041, 5.3.2(a)(b) and 5.3.3(a)(b)
IFSTA, *Fire and Emergency Services Instructor,* 6th Edition, 4th Printing, page 126.
Answer: A

41. Reference: NFPA 1041, 5.3.2(a)(b), 5.3.3(a)(b), and 5.3.1
IFSTA, *Fire and Emergency Services Instructor,* 6th Edition, 4th Printing, pages 116–117.
Answer: A

42. Reference: NFPA 1041, 5.3.3(a)(b) and 5.3.2(a)
IFSTA, *Fire and Emergency Services Instructor,* 6th Edition, 4th Printing, pages 126–127.
Answer: B

43. Reference: NFPA 1041, 5.3.3(a)(b)
IFSTA, *Fire and Emergency Services Instructor,* 6th Edition, 4th Printing, page 81.
Answer: A

44. Reference: NFPA 1041, 5.4.2(a)(b), 5.3.2(a)(b), and 5.3.3(a)
IFSTA, *Fire and Emergency Services Instructor,* 6th Edition, 4th Printing, page 122.
Answer: B

45. Reference: NFPA 1041, 5.4.2(a)(b)
IFSTA, *Fire and Emergency Services Instructor,* 6th Edition, 4th Printing, page 137.
Answer: B

46. Reference: NFPA 1041, 5.4.2(a)(b)
IFSTA, *Fire and Emergency Services Instructor,* 6th Edition, 4th Printing, page 142.
Answer: C

47. Reference: NFPA 1041, 5.4.2(a)(b)
IFSTA, *Fire and Emergency Services Instructor,* 6th Edition, 4th Printing, pages 163–164.
Answer: D

48. Reference: NFPA 1041, 5.4.2(a)(b)
IFSTA, *Fire and Emergency Services Instructor,* 6th Edition, 4th Printing, pages 162–163.
Answer: B

49. Reference: NFPA 1041, 5.4.2(a)(b), and 5.3.2(a)(b)
IFSTA, *Fire and Emergency Services Instructor,* 6th Edition, 4th Printing, page 164.
Answer: B

50. Reference: NFPA 1041, 5.4.2(a)(b)
IFSTA, *Fire and Emergency Services Instructor,* 6th Edition, 4th Printing, page 9.
Answer: B

51. Reference: NFPA 1041, 5.4.2(a)(b) and 5.4.1
IFSTA, *Fire and Emergency Services Instructor,* 6th Edition, 4th Printing, page 117.
Answer: B

52. Reference: NFPA 1041, 5.4.3(a)(b)
IFSTA, *Fire and Emergency Services Instructor,* 6th Edition, 4th Printing, pages 171–172.
Answer: B

53. Reference: NFPA 1041, 5.4.3(a)
IFSTA, *Fire and Emergency Services Instructor,* 6th Edition, 4th Printing, pages 37.
Answer: D

54. Reference: NFPA 1041, 5.4.3(a)
IFSTA, *Fire and Emergency Services Instructor,* 6th Edition, 4th Printing, pages 44–45.
Answer: B

55. Reference: NFPA 1041, 5.4.3(a)
IFSTA, *Fire and Emergency Services Instructor,* 6th Edition, 4th Printing, page 171.
Answer: D

56. Reference: NFPA 1041, 5.4.3(a)(b)
IFSTA, *Fire and Emergency Services Instructor,* 6th Edition, 4th Printing, pages 177–178.
Answer: A

57. Reference: NFPA 1041, 5.4.3(a)
IFSTA, *Fire and Emergency Services Instructor,* 6th Edition, 4th Printing, page 185.
Answer: D

58. Reference: NFPA 1041, 5.4.3(a)
IFSTA, *Fire and Emergency Services Instructor,* 6th Edition, 4th Printing, page 238.
Answer: D

59. Reference: NFPA 1041, 5.4.3(a)(b)
IFSTA, *Fire and Emergency Services Instructor,* 6th Edition, 4th Printing, pages 145–146.
Answer: C

60. Reference: NFPA 1041, 5.5.2(a)(b)
IFSTA, *Fire and Emergency Services Instructor,* 6th Edition, 4th Printing, page 191.
Answer: A

61. Reference: NFPA 1041, 5.5.2(a)(b)
IFSTA, *Fire and Emergency Services Instructor,* 6th Edition, 4th Printing, page 193.
Answer: B

62. Reference: NFPA 1041, 5.5.2(a)(b)
IFSTA, *Fire and Emergency Services Instructor,* 6th Edition, 4th Printing, page 191.
Answer: A

63. Reference: NFPA 1041, 5.5.2(a)(b)
IFSTA, *Fire and Emergency Services Instructor,* 6th Edition, 4th Printing, page 205.
Answer: C

64. Reference: NFPA 1041, 5.5.2(a)(b)
IFSTA, *Fire and Emergency Services Instructor,* 6th Edition, 4th Printing, page 192.
Answer: A

65. Reference: NFPA 1041, 5.5.2(a)(b)
IFSTA, *Fire and Emergency Services Instructor,* 6th Edition, 4th Printing, page 203.
Answer: D

66. Reference: NFPA 1041, 5.5.2(a)(b), 5.5.3(a), and 5.5.1
IFSTA, *Fire and Emergency Services Instructor,* 6th Edition, 4th Printing, page 199.
Answer: D

67. Reference: NFPA 1041, 5.5.2(a)(b) and 5.4.2(a)(b)
IFSTA, *Fire and Emergency Services Instructor,* 6th Edition, 4th Printing, page 146.
Answer: B

68. Reference: NFPA 1041, 5.5.3(a)(b)
IFSTA, *Fire and Emergency Services Instructor,* 6th Edition, 4th Printing, pages 216–217.
Answer: A

69. Reference: NFPA 1041, 5.5.3(a)(b) and 5.2.6(a)(b)
IFSTA, *Fire and Emergency Services Instructor,* 6th Edition, 4th Printing, pages 215–216.
Answer: A

70. Reference: NFPA 1041, 5.5.3(a)(b), 5.5.2(a)(b), and 5.3.3(a)(b)
IFSTA, *Fire and Emergency Services Instructor,* 6th Edition, 4th Printing, page 102.
Answer: A

71. Reference: NFPA 1041, 5.5.3(a)(b), 5.5.2(a)(b), 5.3.3(a)(b), and 5.5.4(a)(b)
IFSTA, *Fire and Emergency Services Instructor,* 6th Edition, 4th Printing, page 106.
Answer: A

72. Reference: NFPA 1041, 5.5.4(a) and 5.5.2(a)(b)
IFSTA, *Fire and Emergency Services Instructor,* 6th Edition, 4th Printing, page 199.
Answer: B

73. Reference: NFPA 1041, 5.5.4(a) and 5.5.2(a)(b)
IFSTA, *Fire and Emergency Services Instructor,* 6th Edition, 4th Printing, page 192.
Answer: B

74. Reference: NFPA 1041, 5.5.4(a)(b) and 5.5.2(a)(b)
IFSTA, *Fire and Emergency Services Instructor,* 6th Edition, 4th Printing, page 213.
Answer: D

75. Reference: NFPA 1041, 5.5.4(a)
IFSTA, *Fire and Emergency Services Instructor,* 6th Edition, 4th Printing, pages 214–215.
Answer: A

Don't forget to enter the information on your Personal Progress Plotter and answer the Yes and No question at the end of the Examination. This step is extremely important for the successful completion of the Systematic Approach to Examination Preparation!

Examination II-2 Answer Key

Directions

Follow these steps carefully for completing the feedback part of SAEP:

1. After calculating your score, look up the answers for the examination items you missed as well as those on which you guessed, even if you guessed correctly. If you are guessing, it means the answer is not perfectly clear. In this process, we are committed to making you as knowledgeable as possible.
2. Enter the number of missed and guessed examination items in the blanks on your Personal Progress Plotter.
3. Highlight the answer in the reference materials. Read the paragraph preceding and the paragraph following the one in which the correct answer is located. Enter the paragraph number and page number next to the guessed or missed examination item on your examination. Count any part of a paragraph at the beginning of the page as one paragraph until you reach the paragraph containing your highlighted answer. This step will help you locate and review your missed and guessed examination items later in the process. It is essential to learning the material in context and by association. These learning techniques (context/association) are the very backbone of the SAEP approach.
4. Once you have completed the feedback step, you may proceed to the next examination.

1. Reference: NFPA 1041, 5.2.2(a)
IFSTA, *Fire and Emergency Services Instructor,* 6th Edition, 4th Printing, page 258.
Answer: C

2. Reference: NFPA 1041, 5.2.2(a), 5.2.1, and 3.3.2.2
IFSTA, *Fire and Emergency Services Instructor,* 6th Edition, 4th Printing, pages 3–4.
Answer: C

3. Reference: NFPA 1041, 5.2.2(a)
IFSTA, *Fire and Emergency Services Instructor,* 6th Edition, 4th Printing, pages 106–109.
Answer: B

4. Reference: NFPA 1041, 5.2.2(a)
IFSTA, *Fire and Emergency Services Instructor,* 6th Edition, 4th Printing, pages 107.
Answer: E

5. Reference: NFPA 1041, 5.2.3(a)(b)
IFSTA, *Fire and Emergency Services Instructor,* 6th Edition, 4th Printing, page 92.
Answer: B

6. Reference: NFPA 1041, 5.2.3(a)(b) and 5.2.4(a)(b)
IFSTA, *Fire and Emergency Services Instructor,* 6th Edition, 4th Printing, pages 262–263.
Answer: B

7. Reference: NFPA 1041, 5.2.3(a)(b) and 5.2.4(a)(b)
IFSTA, *Fire Department Company Officer*, 3rd Edition, 1st Printing, pages 153–154.
Answer: C

8. Reference: NFPA 1041, 5.2.3(a)(b) and 5.2.4(a)(b)
IFSTA, *Fire and Emergency Services Instructor,* 6th Edition, 4th Printing, page 263.
Answer: D

9. Reference: NFPA 1041, 5.2.4(a) and 5.2.3(a)(b)
IFSTA, *Fire and Emergency Services Instructor,* 6th Edition, 4th Printing, page 262.
Answer: B

10. Reference: NFPA 1041, 5.2.4(a) and 5.2.3(a)(b)
IFSTA, *Fire and Emergency Services Instructor,* 6th Edition, 4th Printing, page 262.
Answer: A

11. Reference: NFPA 1041, 5.2.4(a) and 5.2.3(a)(b)
IFSTA, *Fire and Emergency Services Instructor,* 6th Edition, 4th Printing, page 262.
Answer: D

12. Reference: NFPA 1041, 5.2.5(a)(b)
IFSTA, *Fire and Emergency Services Instructor,* 6th Edition, 4th Printing, page 39.
Answer: B

13. Reference: NFPA 1041, 5.2.5(a)(b)
IFSTA, *Fire and Emergency Services Instructor,* 6th Edition, 4th Printing, page 39.
Answer: D

14. Reference: NFPA 1041, 5.2.5(a), 5.2.6(a), and 5.4.3(a)
IFSTA, *Fire and Emergency Services Instructor,* 6th Edition, 4th Printing, page 248.
Answer: B

15. Reference: NFPA 1041, 5.2.5(a)(b)
IFSTA, *Fire and Emergency Services Instructor,* 6th Edition, 4th Printing, pages 38–39.
Answer: B

16. Reference: NFPA 1041, 5.2.5(a)(b)
IFSTA, *Fire and Emergency Services Instructor,* 6th Edition, 4th Printing, page 39.
Answer: C

17. Reference: NFPA 1041, 5.2.6(a)(b)
IFSTA, *Fire and Emergency Services Instructor,* 6th Edition, 4th Printing, pages 18–19.
Answer: A

18. Reference: NFPA 1041, 5.2.6(a)(b)
IFSTA, *Fire and Emergency Services Instructor,* 6th Edition, 4th Printing, page 99.
Answer: C

19. Reference: NFPA 1041, 5.2.6(a)(b)
IFSTA, *Fire and Emergency Services Instructor,* 6th Edition, 4th Printing, page 99.
Answer: D

20. Reference: NFPA 1041, 5.2.6(a)(b)
IFSTA, *Fire and Emergency Services Instructor,* 6th Edition, 4th Printing, page 260.
Answer: D

21. Reference: NFPA 1041, 5.3.2, 5.3.2(a)(b), 5.3.3, and 5.3.3(a)(b)
IFSTA, *Fire and Emergency Services Instructor,* 6th Edition, 4th Printing, pages 119 and 126-127.
Answer: B

22. Reference: NFPA 1041, 5.3.2 and 5.3.2(a)(b)
IFSTA, *Fire and Emergency Services Instructor,* 6th Edition, 4th Printing, page 95.
Answer: C

23. Reference: NFPA 1041, 5.3.2, 5.3.2(a)(b), and 5.3.3(a)(b)
IFSTA, *Fire and Emergency Services Instructor,* 6th Edition, 4th Printing, pages 103–104.
Answer: C

24. Reference: NFPA 1041, 5.3.2(a)(b) and 5.3.1
IFSTA, *Fire and Emergency Services Instructor,* 6th Edition, 4th Printing, pages 245–246.
Answer: D

25. Reference: NFPA 1041, 5.3.2, 5.3.2(a)(b), and 5.3.3(a)(b)
IFSTA, *Fire and Emergency Services Instructor,* 6th Edition, 4th Printing, page 103.
Answer: B

26. Reference: NFPA 1041, 5.3.2, 5.3.2(a)(b), and 5.3.3(a)(b)
IFSTA, *Fire and Emergency Services Instructor,* 6th Edition, 4th Printing, pages 119 and 126–127.
Answer: B

27. Reference: NFPA 1041, 5.3.2, and 5.3.2 (a)(b)
IFSTA, *Fire and Emergency Services Instructor,* 6th Edition, 4th Printing, page 104.
Answer: C

28. Reference: NFPA 1041, 5.3.2, 5.3.2(a)(b), and 5.3.3(a)(b)
IFSTA, *Fire and Emergency Services Instructor,* 6th Edition, 4th Printing, page 125.
Answer: A

29. Reference: NFPA 1041, 5.3.2(a)(b) and 5.3.3(a)(b)
IFSTA, *Fire and Emergency Services Instructor,* 6th Edition, 4th Printing, page 226.
Answer: B

30. Reference: NFPA 1041, 5.3.2, 5.3.2(a)(b), and 5.3.3(a)(b)
IFSTA, *Fire and Emergency Services Instructor,* 6th Edition, 4th Printing, page 238.
Answer: D

31. Reference: NFPA 1041, 5.3.2, 5.3.2(a)(b), 5.3.3, and 5.3.3(a)(b)
IFSTA, *Fire and Emergency Services Instructor,* 6th Edition, 4th Printing, pages 49–50.
Answer: C

32. Reference: NFPA 1041, 5.3.2(a)(b) and 5.3.3(a)(b)
IFSTA, *Fire and Emergency Services Instructor,* 6th Edition, 4th Printing, page 58.
Answer: D

33. Reference: NFPA 1041, 5.3.2(a)(b) and 5.3.3(a)(b)
IFSTA, *Fire and Emergency Services Instructor,* 6th Edition, 4th Printing, page 100.
Answer: D

34. Reference: NFPA 1041, 5.3.2(a)(b) and 5.3.3(a)(b)
IFSTA, *Fire and Emergency Services Instructor,* 6th Edition, 4th Printing, page 100.
Answer: A

35. Reference: NFPA 1041, 5.3.2(a)(b) and 5.3.3(a)(b)
IFSTA, *Fire and Emergency Services Instructor,* 6th Edition, 4th Printing, page 102.
Answer: A

36. Reference: NFPA 1041, 5.3.2(a)(b) and 5.3.3(a)(b)
IFSTA, *Fire and Emergency Services Instructor,* 6th Edition, 4th Printing, page 114.
Answer: D

37. Reference: NFPA 1041, 5.3.2(a)(b) and 5.3.3(a)(b)
IFSTA, *Fire and Emergency Services Instructor,* 6th Edition, 4th Printing, page 124.
Answer: A

38. Reference: NFPA 1041, 5.3.2(a)(b) and 5.3.3(a)(b)
IFSTA, *Fire and Emergency Services Instructor,* 6th Edition, 4th Printing, page 104.
Answer: C

39. Reference: NFPA 1041, 5.3.2(a)(b) and 5.3.3(a)(b)
IFSTA, *Fire and Emergency Services Instructor,* 6th Edition, 4th Printing, page 104.
Answer: B

40. Reference: NFPA 1041, 5.3.2(a)(b) and 5.3.3(a)(b)
IFSTA, *Fire and Emergency Services Instructor,* 6th Edition, 4th Printing, page 104.
Answer: D

41. Reference: NFPA 1041, 5.3.2 and 5.3.2.(a)(b)
IFSTA, *Fire and Emergency Services Instructor,* 6th Edition, 4th Printing, pages 95, 102, 103, and 104.
Answer: C

42. Reference: NFPA 1041, 5.3.2(a)(b), 5.3.3(a)(b), and 5.4.1
IFSTA, *Fire and Emergency Services Instructor,* 6th Edition, 4th Printing, page 114.
Answer: D

43. Reference: NFPA 1041, 5.3.2(a)(b) and 5.3.3(a)(b)
IFSTA, *Fire and Emergency Services Instructor,* 6th Edition, 4th Printing, page 53.
Answer: D

44. Reference: NFPA 1041, 5.4.2(a)
IFSTA, *Fire and Emergency Services Instructor,* 6th Edition, 4th Printing, page 136.
Answer: D

45. Reference: NFPA 1041, 5.4.2(a)(b) and 5.3.3(a)(b)
IFSTA, *Fire and Emergency Services Instructor,* 6th Edition, 4th Printing, page 162.
Answer: A

46. Reference: NFPA 1041, 5.4.2(a)(b)
IFSTA, *Fire and Emergency Services Instructor,* 6th Edition, 4th Printing, pages 163–164.
Answer: D

47. Reference: NFPA 1041, 5.4.2 (a)(b) and 5.3.3.(a)(b)
IFSTA, *Fire and Emergency Services Instructor,* 6th Edition, 4th Printing, page 164.
Answer: C

48. Reference: NFPA 1041, 5.4.2(a)(b), 5.3.2(a)(b), 5.3.3(a)(b), 5.3.1, and 5.4.1
IFSTA, *Fire and Emergency Services Instructor,* 6th Edition, 4th Printing, page 120.
Answer: A

49. Reference: NFPA 1041, 5.4.2(a)(b)
IFSTA, *Fire and Emergency Services Instructor,* 6th Edition, 4th Printing, page 83.
Answer: A

50. Reference: NFPA 1041, 5.4.2(a)(b)
IFSTA, *Fire and Emergency Services Instructor,* 6th Edition, 4th Printing, page 9.
Answer: B

51. Reference: NFPA 1041, 5.4.2(a)(b), 5.3.1, 5.3.2(a)(b), 5.3.3(a)(b), and 5.4.1
IFSTA, *Fire and Emergency Services Instructor,* 6th Edition, 4th Printing, page 166.
Answer: D

52. Reference: NFPA 1041, 5.4.3(a)(b)
IFSTA, *Fire and Emergency Services Instructor,* 6th Edition, 4th Printing, page 164.
Answer: D

53. Reference: NFPA 1041, 5.4.3(a)
IFSTA, *Fire and Emergency Services Instructor,* 6th Edition, 4th Printing, page 25.
Answer: B

54. Reference: NFPA 1041, 5.4.3(a)
IFSTA, *Fire and Emergency Services Instructor,* 6th Edition, 4th Printing, pages 26–27.
Answer: D

55. Reference: NFPA 1041, 5.4.3(a)
IFSTA, *Fire and Emergency Services Instructor,* 6th Edition, 4th Printing, page 29.
Answer: A

56. Reference: NFPA 1041, 5.4.3(a)
IFSTA, *Fire and Emergency Services Instructor,* 6th Edition, 4th Printing, page 32.
Answer: C

57. Reference: NFPA 1041, 5.4.3(a)
IFSTA, *Fire and Emergency Services Instructor,* 6th Edition, 4th Printing, page 45.
Answer: B

58. Reference: NFPA 1041, 5.4.3(a)
IFSTA, *Fire and Emergency Services Instructor,* 6th Edition, 4th Printing, page 146.
Answer: B

59. Reference: NFPA 1041, 5.4.3(a)(b)
IFSTA, *Fire and Emergency Services Instructor,* 6th Edition, 4th Printing, pages 145–146.
Answer: C

60. Reference: NFPA 1041, 5.5.2(a)(b)
IFSTA, *Fire and Emergency Services Instructor,* 6th Edition, 4th Printing, pages 193 and 200.
Answer: D

61. Reference: NFPA 1041, 5.5.2(a)(b)
IFSTA, *Fire and Emergency Services Instructor,* 6th Edition, 4th Printing, page 203.
Answer: B

62. Reference: NFPA 1041, 5.5.2(a)(b)
IFSTA, *Fire and Emergency Services Instructor,* 6th Edition, 4th Printing, page 192.
Answer: C

63. Reference: NFPA 1041, 5.5.2(a)(b)
IFSTA, *Fire and Emergency Services Instructor,* 6th Edition, 4th Printing, page 208.
Answer: A

64. Reference: NFPA 1041, 5.5.2(a)(b)
IFSTA, *Fire and Emergency Services Instructor,* 6th Edition, 4th Printing, page 208.
Answer: A

65. Reference: NFPA 1041, 5.5.2(a)(b)
IFSTA, *Fire and Emergency Services Instructor,* 6th Edition, 4th Printing, page 206–208.
Answer: D

66. Reference: NFPA 1041, 5.5.2(a)(b)
IFSTA, *Fire and Emergency Services Instructor,* 6th Edition, 4th Printing, page 193.
Answer: C

67. Reference: NFPA 1041, 5.5.2(a)(b), 5.5.1, and 5.5.3(a)
IFSTA, *Fire and Emergency Services Instructor,* 6th Edition, 4th Printing, page 199.
Answer: B

68. Reference: NFPA 1041, 5.5.2(a)(b), 5.5.3(a), and 5.5.1
IFSTA, *Fire and Emergency Services Instructor,* 6th Edition, 4th Printing, page 199.
Answer: D

69. Reference: NFPA 1041, 5.53(a)(b), 5.5.2(a)(b), and 5.3.3(a)(b)
IFSTA, *Fire and Emergency Services Instructor,* 6th Edition, 4th Printing, pages 217–218.
Answer: A

70. Reference: NFPA 1041, 5.5.3(a)(b), 5.5.2(a)(b), and 5.3.3(a)(b)
IFSTA, *Fire and Emergency Services Instructor,* 6th Edition, 4th Printing, page 217.
Answer: C

71. Reference: NFPA 1041, 5.5.3(a)(b), 5.5.2(a)(b), and 5.3.3(a)(b)
IFSTA, *Fire and Emergency Services Instructor,* 6th Edition, 4th Printing, page 217.
Answer: D

72. Reference: NFPA 1041, 5.5.3(a)(b), 5.5.2(a)(b), and 5.3.3(a)(b)
IFSTA, *Fire and Emergency Services Instructor,* 6th Edition, 4th Printing, pages 215–216.
Answer: A

73. Reference: NFPA 1041, 5.5.4(a)(b)
IFSTA, *Fire and Emergency Services Instructor,* 6th Edition, 4th Printing, page 214.
Answer: B

74. Reference: NFPA 1041, 5.5.4(a)(b), 5.5.3(a), 5.5.2(a)(b), 5.3.3(a)(b), and 5.3.2(a)(b)
IFSTA, *Fire and Emergency Services Instructor,* 6th Edition, 4th Printing, page 106.
Answer: B

75. Reference: NFPA 1041, 5.5.4(a)
IFSTA, *Fire and Emergency Services Instructor,* 6th Edition, 4th Printing, page 214.
Answer: A

Don't forget to enter the information on your Personal Progress Plotter and answer the Yes and No question at the end of the Examination. This step is extremely important for the successful completion of the Systematic Approach to Examination Preparation!

Examination II-3 Answer Key

Directions

Follow these steps carefully for completing the feedback part of SAEP:

1. After calculating your score, look up the answers for the examination items you missed as well as those on which you guessed, even if you guessed correctly. If you are guessing, it means the answer is not perfectly clear. In this process, we are committed to making you as knowledgeable as possible.
2. Enter the number of missed and guessed examination items in the blanks on your Personal Progress Plotter.
3. Highlight the answer in the reference materials. Read the paragraph preceding and the paragraph following the one in which the correct answer is located. Enter the paragraph number and page number next to the guessed or missed examination item on your examination. Count any part of a paragraph at the beginning of the page as one paragraph until you reach the paragraph containing your highlighted answer. This step will help you locate and review your missed and guessed examination items later in the process. It is essential to learning the material in context and by association. These learning techniques (context/association) are the very backbone of the SAEP approach.
4. Congratulations! You have completed the examination and feedback parts of SAEP when you have highlighted your guessed and missed examination items for this examination.

Proceed to Phases III and IV. Study the materials carefully in these important phases-they will help you polish your examination-taking skills. Approximately two to three days before you take your next examination, carefully read all of the highlighted information in the reference materials using the same techniques you applied during the feedback part. This will reinforce your learning and provide you with an added level of confidence going into the examination.

Someone once said to professional golfer Tom Watson after he won several tournament championships, "You are really lucky to have won those championships. You are really on a streak." Watson was reported to have replied, "Yes, there is some luck involved, but what I've really noticed is that the more I practice, the luckier I get." What Watson was saying is that good luck usually results from good preparation. This line of thinking certainly applies to learning the rules and hints of examination taking.

Rule 7

Good luck = good preparation.

---

1. Reference: NFPA 1041, 5.2.2(a) and 5.2.1
   IFSTA, *Fire and Emergency Services Instructor,* 6th Edition, 4th Printing, page 66.
   Answer: A

2. Reference: NFPA 1041, 5.2.2(a)
   IFSTA, *Fire and Emergency Services Instructor,* 6th Edition, 4th Printing, page 258.
   Answer: C

3. Reference: NFPA 1041, 5.2.2(a), 5.2.1, and 3.3.2.2
IFSTA, *Fire and Emergency Services Instructor,* 6th Edition, 4th Printing, pages 3–4.
Answer: C

4. Reference: NFPA 1041, 5.2.2(a) and 5.2.1
IFSTA, *Fire and Emergency Services Instructor,* 6th Edition, 4th Printing, pages 18–19.
Answer: D

5. Reference: NFPA 1041, 5.2.2(a)
IFSTA, *Fire and Emergency Services Instructor,* 6th Edition, 4th Printing, pages 107.
Answer: A

6. Reference: NFPA 1041, 5.2.2(a), 5.2.1, and 5.2.5(a)
IFSTA, *Fire and Emergency Services Instructor,* 6th Edition, 4th Printing, page 259.
Answer: A

7. Reference: NFPA 1041, 5.2.2(a)
IFSTA, *Fire and Emergency Services Instructor,* 6th Edition, 4th Printing, pages 106–109.
Answer: C

8. Reference: NFPA 1041, 5.2.2(a)
IFSTA, *Fire and Emergency Services Instructor,* 6th Edition, 4th Printing, pages 107.
Answer: E

9. Reference: NFPA 1041, 5.2.3(a)(b) and 5.2.4(a)(b)
IFSTA, *Fire and Emergency Services Instructor,* 6th Edition, 4th Printing, page 262.
Answer: A

10. Reference: NFPA 1041, 5.2.3(a)(b) and 5.2.4(a)(b)
IFSTA, *Fire and Emergency Services Instructor,* 6th Edition, 4th Printing, page 263.
Answer: D

11. Reference: NFPA 1041, 5.2.4(a)
IFSTA, *Fire and Emergency Services Instructor,* 6th Edition, 4th Printing, page 262.
Answer: B

12. Reference: NFPA 1041, 5.2.4(a) and 5.2.3(a)(b)
IFSTA, *Fire and Emergency Services Instructor,* 6th Edition, 4th Printing, page 262.
Answer: B

13. Reference: NFPA 1041, 5.2.4(a) and 5.2.3(a)(b)
IFSTA, *Fire and Emergency Services Instructor,* 6th Edition, 4th Printing, page 262.
Answer: A

14. Reference: NFPA 1041, 5.2.4(a) and 5.2.3(a)(b)
IFSTA, *Fire and Emergency Services Instructor,* 6th Edition, 4th Printing, page 246.
Answer: D

15. Reference: NFPA 1041, 5.2.4(a)(b) and 5.2.3(a)(b)
IFSTA, *Fire and Emergency Services Instructor,* 6th Edition, 4th Printing, page 262.
Answer: A

16. Reference: NFPA 1041, 5.2.5(a) and 5.2.2(a)
IFSTA, *Fire and Emergency Services Instructor,* 6th Edition, 4th Printing, page 44.
Answer: D

17. Reference: NFPA 1041, 5.2.5(a), 5.2.6(a), and 5.4.3(a)
IFSTA, *Fire and Emergency Services Instructor,* 6th Edition, 4th Printing, page 248.
Answer: B

18. Reference: NFPA 1041, 5.2.5(a)
IFSTA, *Fire and Emergency Services Instructor,* 6th Edition, 4th Printing, page 44.
Answer: C

19. Reference: NFPA 1041, 5.2.5(a)
IFSTA, *Fire and Emergency Services Instructor,* 6th Edition, 4th Printing, page 44.
Answer: B

20. Reference: NFPA 1041, 5.2.5(a)(b)
IFSTA, *Fire and Emergency Services Instructor,* 6th Edition, 4th Printing, page 39.
Answer: C

21. Reference: NFPA 1041, 5.2.6(a)(b)
IFSTA, *Fire and Emergency Services Instructor,* 6th Edition, 4th Printing, pages 257–258.
Answer: B

22. Reference: NFPA 1041, 5.2.6(a)(b)
IFSTA, *Fire and Emergency Services Instructor,* 6th Edition, 4th Printing, page 99.
Answer: C

23. Reference: NFPA 1041, 5.2.6(a)(b)
IFSTA, *Fire and Emergency Services Instructor,* 6th Edition, 4th Printing, page 260.
Answer: D

24. Reference: NFPA 1041, 5.2.6(a)(b)
IFSTA, *Fire and Emergency Services Instructor,* 6th Edition, 4th Printing, page 260.
Answer: A

25. Reference: NFPA 1041, 5.2.6(a)(b), 5.2.3(a)(b), and 5.2.4(a)(b)
IFSTA, *Fire and Emergency Services Instructor,* 6th Edition, 4th Printing, page 260.
Answer: C

26. Reference: NFPA 1041, 5.2.6(a)(b)
IFSTA, *Fire and Emergency Services Instructor,* 6th Edition, 4th Printing, pages 165–166.
Answer: A

27. Reference: NFPA 1041, 5.3.2(a)(b) and 5.3.3(a)(b)
IFSTA, *Fire and Emergency Services Instructor,* 6th Edition, 4th Printing, page 95.
Answer: D

28. Reference: NFPA 1041, 5.3.2(a)(b) and 5.3.3(a)(b)
IFSTA, *Fire and Emergency Services Instructor,* 6th Edition, 4th Printing, pages 102–103.
Answer: C

29. Reference: NFPA 1041, 5.3.2, 5.3.2(a)(b), 5.3.3, and 5.3.3(a)(b)
IFSTA, *Fire and Emergency Services Instructor,* 6th Edition, 4th Printing, pages 119 and 126–127.
Answer: B

30. Reference: NFPA 1041, 5.3.2, 5.3.2(a)(b), and 5.3.3(a)(b)
IFSTA, *Fire and Emergency Services Instructor,* 6th Edition, 4th Printing, pages 103–104.
Answer: C

31. Reference: NFPA 1041, 5.3.2, 5.3.2(a)(b), and 5.3.3(a)(b)
IFSTA, *Fire and Emergency Services Instructor,* 6th Edition, 4th Printing, page 103.
Answer: D

32. Reference: NFPA 1041, 5.3.2, 5.3.2(a)(b), and 5.3.3(a)(b)
IFSTA, *Fire and Emergency Services Instructor,* 6th Edition, 4th Printing, pages 102 and 194.
Answer: D

33. Reference: NFPA 1041, 5.3.2, 5.3.2(a)(b), and 5.3.3(a)(b)
IFSTA, *Fire and Emergency Services Instructor,* 6th Edition, 4th Printing, pages 119 and 126–127.
Answer: B

34. Reference: NFPA 1041, 5.3.2 and 5.3.2 (a)(b)
IFSTA, *Fire and Emergency Services Instructor,* 6th Edition, 4th Printing, page 104.
Answer: C

35. Reference: NFPA 1041, 5.3.2, 5.3.2.(a)(b), and 5.3.3(a)(b)
IFSTA, *Fire and Emergency Services Instructor,* 6th Edition, 4th Printing, pages 238–239.
Answer: A

36. Reference: NFPA 1041, 5.3.2, 5.3.2(a)(b), and 5.3.3(a)(b)
IFSTA, *Fire and Emergency Services Instructor,* 6th Edition, 4th Printing, page 125.
Answer: A

37. Reference: NFPA 1041, 5.3.2, 5.3.2(a)(b), and 5.3.3(a)(b)
IFSTA, *Fire and Emergency Services Instructor,* 6th Edition, 4th Printing, pages 236–237.
Answer: D

38. Reference: NFPA 1041, 5.3.2, 5.3.2(a)(b), 5.3.3, and 5.3.3(a)(b)
IFSTA, *Fire and Emergency Services Instructor,* 6th Edition, 4th Printing, page 226.
Answer: B

39. Reference: NFPA 1041, 5.3.2(a)(b) and 5.3.3(a)(b)
IFSTA, *Fire and Emergency Services Instructor,* 6th Edition, 4th Printing, page 104.
Answer: A

40. Reference: NFPA 1041, 5.3.2, 5.3.2(a)(b), 5.3.3, and 5.3.3(a)(b)
IFSTA, *Fire and Emergency Services Instructor,* 6th Edition, 4th Printing, page 98.
Answer: B

41. Reference: NFPA 1041, 5.3.2(a)(b) and 5.3.3(a)(b)
IFSTA, *Fire and Emergency Services Instructor,* 6th Edition, 4th Printing, page 58.
Answer: C

42. Reference: NFPA 1041, 5.3.2(a)(b) and 5.3.3(a)(b)
IFSTA, *Fire and Emergency Services Instructor,* 6th Edition, 4th Printing, page 58.
Answer: D

43. Reference: NFPA 1041, 5.3.2(a)(b) and 5.3.3(a)(b)
IFSTA, *Fire and Emergency Services Instructor,* 6th Edition, 4th Printing, page 61.
Answer: C

44. Reference: NFPA 1041, 5.3.2(a)(b) and 5.3.3(a)(b)
IFSTA, *Fire and Emergency Services Instructor,* 6th Edition, 4th Printing, pages 93–94.
Answer: D

45. Reference: NFPA 1041, 5.3.2(a)(b) and 5.3.3(a)(b)
IFSTA, *Fire and Emergency Services Instructor,* 6th Edition, 4th Printing, page 100.
Answer: D

46. Reference: NFPA 1041, 5.3.2(a)(b) and 5.3.3(a)(b)
IFSTA, *Fire and Emergency Services Instructor,* 6th Edition, 4th Printing, page 102.
Answer: A

47. Reference: NFPA 1041, 5.3.2(a)(b) and 5.3.3(a)(b)
IFSTA, *Fire and Emergency Services Instructor,* 6th Edition, 4th Printing, page 104.
Answer: D

48. Reference: NFPA 1041, 5.3.2(a)(b) and 5.3.3(a)(b)
IFSTA, *Fire and Emergency Services Instructor,* 6th Edition, 4th Printing, page 229.
Answer: B

49. Reference: NFPA 1041, 5.3.2 and 5.3.2(a)(b)
IFSTA, *Fire and Emergency Services Instructor,* 6th Edition, 4th Printing, page 55.
Answer: A

50. Reference: NFPA 1041, 5.3.2(a)(b) and 5.3.3(a)(b)
IFSTA, *Fire and Emergency Services Instructor,* 6th Edition, 4th Printing, page 119.
Answer: D

51. Reference: NFPA 1041, 5.3.2(a)(b), 5.3.3(a)(b), and 5.4.2(a)(b)
IFSTA, *Fire and Emergency Services Instructor,* 6th Edition, 4th Printing, page 70–72.
Answer: D

52. Reference: NFPA 1041, 5.3.2(a)(b), 5.3.3(a)(b), and 5.4.2(a)(b)
IFSTA, *Fire and Emergency Services Instructor,* 6th Edition, 4th Printing, page 77.
Answer: D

53. Reference: NFPA 1041, 5.3.2(a)(b), 5.3.3(a)(b), and 5.4.2(a)(b)
IFSTA, *Fire and Emergency Services Instructor,* 6th Edition, 4th Printing, page 74.
Answer: B

54. Reference: NFPA 1041, 5.3.2(a) and 5.3.3(a)
IFSTA, *Fire and Emergency Services Instructor,* 6th Edition, 4th Printing, pages 116–118.
Answer: D

55. Reference: NFPA 1041, 5.3.2(a)(b) and 5.3.3(a)(b)
IFSTA, *Fire and Emergency Services Instructor,* 6th Edition, 4th Printing, page 53.
Answer: D

56. Reference: NFPA 1041, 5.3.3(a)(b) and 5.3.2(a)
IFSTA, *Fire and Emergency Services Instructor,* 6th Edition, 4th Printing, pages 126–127.
Answer: B

57. Reference: NFPA 1041, 5.4.2(a)
IFSTA, *Fire and Emergency Services Instructor,* 6th Edition, 4th Printing, page 136.
Answer: D

58. Reference: NFPA 1041, 5.4.2(a)(b)
IFSTA, *Fire and Emergency Services Instructor,* 6th Edition, 4th Printing, page 137.
Answer: B

59. Reference: NFPA 1041, 5.4.2(a)(b)
IFSTA, *Fire and Emergency Services Instructor,* 6th Edition, 4th Printing, pages 163–164.
Answer: D

60. Reference: NFPA 1041, 5.4.2(a)(b)
IFSTA, *Fire and Emergency Services Instructor,* 6th Edition, 4th Printing, pages 159–160.
Answer: B

61. Reference: NFPA 1041, 5.4.2(a)(b) and 5.3.2(a)(b)
IFSTA, *Fire and Emergency Services Instructor,* 6th Edition, 4th Printing, page 164.
Answer: B

62. Reference: NFPA 1041, 5.4.2(a)(b) and 5.3.2(a)(b)
IFSTA, *Fire and Emergency Services Instructor,* 6th Edition, 4th Printing, pages 160–161.
Answer: A

63. Reference: NFPA 1041, 5.4.2(a)(b) and 5.3.2(a)(b)
IFSTA, *Fire and Emergency Services Instructor,* 6th Edition, 4th Printing, page 164.
Answer: B

64. Reference: NFPA 1041, 5.4.2 (a)(b) and 5.3.3.(a)(b)
IFSTA, *Fire and Emergency Services Instructor,* 6th Edition, 4th Printing, page 164.
Answer: C

65. Reference: NFPA 1041, 5.4.2(a)(b), 5.3.2(a)(b), 5.3.3(a)(b), 5.3.1, and 5.4.1
IFSTA, *Fire and Emergency Services Instructor,* 6th Edition, 4th Printing, page 120.
Answer: A

66. Reference: NFPA 1041, 5.4.2(a)(b)
IFSTA, *Fire and Emergency Services Instructor,* 6th Edition, 4th Printing, page 9.
Answer: B

67. Reference: NFPA 1041, 5.4.2(a)(b) and 5.4.1
IFSTA, *Fire and Emergency Services Instructor,* 6th Edition, 4th Printing, pages 145–146.
Answer: D

68. Reference: NFPA 1041, 5.4.2(a)(b) and 5.4.1
IFSTA, *Fire and Emergency Services Instructor,* 6th Edition, 4th Printing, page 117.
Answer: B

69. Reference: NFPA 1041, 5.4.2(a)(b) and 5.4.1
IFSTA, *Fire and Emergency Services Instructor,* 6th Edition, 4th Printing, pages 145–146.
Answer: B

70. Reference: NFPA 1041, 5.4.3(a)
IFSTA, *Fire and Emergency Services Instructor,* 6th Edition, 4th Printing, page 25.
Answer: B

71. Reference: NFPA 1041, 5.4.3(a)
IFSTA, *Fire and Emergency Services Instructor,* 6th Edition, 4th Printing, page 25.
Answer: A

72. Reference: NFPA 1041, 5.4.3(a)
IFSTA, *Fire and Emergency Services Instructor,* 6th Edition, 4th Printing, pages 44–45.
Answer: B

73. Reference: NFPA 1041, 5.4.3(a)
IFSTA, *Fire and Emergency Services Instructor,* 6th Edition, 4th Printing, pages 152–153.
Answer: B

74. Reference: NFPA 1041, 5.4.3(a)(b)
IFSTA, *Fire and Emergency Services Instructor,* 6th Edition, 4th Printing, page 172.
Answer: C

75. Reference: NFPA 1041, 5.4.3(a)
IFSTA, *Fire and Emergency Services Instructor,* 6th Edition, 4th Printing, page 177.
Answer: D

76. Reference: NFPA 1041, 5.4.3(a)(b)
IFSTA, *Fire and Emergency Services Instructor,* 6th Edition, 4th Printing, pages 177–178.
Answer: A

77. Reference: NFPA 1041, 5.4.3(a)(b)
IFSTA, *Fire and Emergency Services Instructor,* 6th Edition, 4th Printing, page 180.
Answer: A

78. Reference: NFPA 1041, 5.4.3(a)
IFSTA, *Fire and Emergency Services Instructor,* 6th Edition, 4th Printing, page 185.
Answer: D

79. Reference: NFPA 1041, 5.5.2(a)(b)
IFSTA, *Fire and Emergency Services Instructor,* 6th Edition, 4th Printing, pages 193 and 200.
Answer: D

80. Reference: NFPA 1041, 5.5.2(a)(b)
IFSTA, *Fire and Emergency Services Instructor,* 6th Edition, 4th Printing, page 203.
Answer: B

81. Reference: NFPA 1041, 5.5.2(a)(b)
IFSTA, *Fire and Emergency Services Instructor,* 6th Edition, 4th Printing, page 192.
Answer: C

82. Reference: NFPA 1041, 5.5.2(a)(b)
IFSTA, *Fire and Emergency Services Instructor,* 6th Edition, 4th Printing, page 192.
Answer: D

83. Reference: NFPA 1041, 5.5.2(a)(b)
IFSTA, *Fire and Emergency Services Instructor,* 6th Edition, 4th Printing, page 193.
Answer: B

84. Reference: NFPA 1041, 5.5.2(a)(b)
IFSTA, *Fire and Emergency Services Instructor,* 6th Edition, 4th Printing, page 208.
Answer: A

85. Reference: NFPA 1041, 5.5.2(a)(b)
IFSTA, *Fire and Emergency Services Instructor,* 6th Edition, 4th Printing, page 208.
Answer: A

86. Reference: NFPA 1041, 5.5.2(a)(b)
IFSTA, *Fire and Emergency Services Instructor,* 6th Edition, 4th Printing, page 191.
Answer: A

87. Reference: NFPA 1041, 5.5.2(a)(b)
IFSTA, *Fire and Emergency Services Instructor,* 6th Edition, 4th Printing, page 192.
Answer: A

88. Reference: NFPA 1041, 5.5.2(a)(b)
IFSTA, *Fire and Emergency Services Instructor,* 6th Edition, 4th Printing, pages 206–208.
Answer: D

89. Reference: NFPA 1041, 5.5.2(a)(b)
IFSTA, *Fire and Emergency Services Instructor,* 6th Edition, 4th Printing, page 203.
Answer: D

90. Reference: NFPA 1041, 5.5.2(a)(b)
IFSTA, *Fire and Emergency Services Instructor,* 6th Edition, 4th Printing, page 192.
Answer: B

91. Reference: NFPA 1041, 5.5.2(a)(b)
IFSTA, *Fire and Emergency Services Instructor,* 6th Edition, 4th Printing, pages 38–39.
Answer: A

92. Reference: NFPA 1041, 5.5.2(a)(b) and 5.5.3(a)(b)
IFSTA, *Fire and Emergency Services Instructor,* 6th Edition, 4th Printing, page 191.
Answer: C

93. Reference: NFPA 1041, 5.5.3(a)(b)
IFSTA, *Fire and Emergency Services Instructor,* 6th Edition, 4th Printing, pages 216–217.
Answer: A

94. Reference: NFPA 1041, 5.5.3(a)(b) and 5.2.6(a)(b)
IFSTA, *Fire and Emergency Services Instructor,* 6th Edition, 4th Printing, pages 215–216.
Answer: A

95. Reference: NFPA 1041, 5.53(a)(b), 5.5.2(a)(b), and 5.3.3(a)(b)
IFSTA, *Fire and Emergency Services Instructor,* 6th Edition, 4th Printing, pages 217–218.
Answer: A

96. Reference: NFPA 1041, 5.5.3(a)(b), 5.5.2(a)(b), and 5.3.3(a)(b)
IFSTA, *Fire and Emergency Services Instructor,* 6th Edition, 4th Printing, page 217.
Answer: C

97. Reference: NFPA 1041, 5.5.3(a)(b) and 5.3.3(a)(b)
IFSTA, *Fire and Emergency Services Instructor,* 6th Edition, 4th Printing, page 216.
Answer: B

98. Reference: NFPA 1041, 5.5.4(a)(b)
IFSTA, *Fire and Emergency Services Instructor,* 6th Edition, 4th Printing, page 214.
Answer: B

99. Reference: NFPA 1041, 5.5.4(a)(b), 5.5.3(a), 5.5.2(a)(b), 5.3.3(a)(b), and 5.3.2(a)(b)
IFSTA, *Fire and Emergency Services Instructor,* 6th Edition, 4th Printing, page 106.
Answer: B

100. Reference: NFPA 1041, 5.5.4(a)
IFSTA, *Fire and Emergency Services Instructor,* 6th Edition, 4th Printing, pages 214–215.
Answer: A

Don't forget to enter the information on your Personal Progress Plotter and answer the Yes and No question at the end of the Examination. This step is extremely important for the successful completion of the Systematic Approach to Examination Preparation!

BIBLIOGRAPHY FOR EXAM PREP: FIRE INSTRUCTOR I AND II

1. NFPA, *Standard for Fire Services Instructor Professional Qualifications*, NFPA 1041, 2002

2. IFSTA, *Fire and Emergency Services Instructor,* Third Edition

3. IFSTA, *Fire Department Company Officer,* Third Edition

Performance Training Systems, Inc.

Training and testing that are on target!

Online examinations for the Fire and Emergency Medical Services

Registration

FREE OFFER - 150 ITEM PRACTICE TEST - VALUED AT $39.00

Complete registration form and fax it to (561) 863-1386.

Name

Title

Department

Address: Street

City State Zip Code

Telephone Fax

E-mail

Choose the tests that apply to your needs.

- ❑ Aerial Operator
- ❑ Airport Fire Fighter
- ❑ Confined Space Rescue
- ❑ EMT Basic
- ❑ Fire and Life Safety Educator I
- ❑ Fire and Life Safety Educator II
- ❑ Fire Inspector 1
- ❑ Fire Inspector 2
- ❑ Fire Instructor 1
- ❑ Fire Instructor 2
- ❑ Fire Investigator
- ❑ Fire Officer 1
- ❑ Fire Officer 2
- ❑ Fire Fighter 1
- ❑ Fire Fighter 2
- ❑ HazMat Awareness and Operations
- ❑ HazMat Technician
- ❑ Pumper Driver
- ❑ Safety Officer
- ❑ Structural Collapse
- ❑ Technical Rescue
- ❑ Telecommunicator I
- ❑ Telecommunicator II
- ❑ Trench Rescue
- ❑ Vehicle/Machinery Rescue
- ❑ Water/Ice Rescue
- ❑ Wildland Fire Fighter I
- ❑ Wildland Fire Fighter II

Signature:______________________________

Made in the USA
Lexington, KY
15 September 2010